OXFORD SURVEYS IN EVOLUTIONARY BIOLOGY

Volume 2
1985

OXFORD SURVEYS IN EVOLUTIONARY BIOLOGY

EDITED BY
R. DAWKINS AND M. RIDLEY

Volume 2
1985

OXFORD UNIVERSITY PRESS
1985

Oxford University Press, Walton Street, OX2 6DP
Oxford New York Toronto
Delhi Bombay Calcutta Madras Karachi
Kuala Lumpur Singapore Hong Kong Tokyo
Nairobi Dar es Salaam Cape Town
Melbourne Auckland
and associated companies in
Beirut Berlin Ibadan Mexico City Nicosia

Oxford is a trade mark of Oxford University Press

Published in the United States
by Oxford University Press, New York

© Oxford University Press, 1985

ISSN 0265–072X

ISBN 0 19 854174 0

Cover illustration from Rudyard Kipling's Just So Stories by kind permission
of The National Trust for Places of Historic Interest or Natural Beauty
and Macmillan (London) Ltd.

Typeset by Oxford Print Associates Ltd
Printed in Great Britain
at the University Press, Oxford
by David Stanford
Printer to the University

Contents

List of contributors

G. C. Williams, Department of Ecology and Evolution, State University of New York at Stony Brook, New York 11794, USA

A. Grafen, Animal Behaviour Research Group, Department of Zoology, South Parks Road, Oxford OX1 3PS, UK

W. J. Sutherland, School of Biological Sciences, University of East Anglia, Norwich NR4 7TJ, UK

H. J. Jerison, Department of Psychiatry and Behavioral Sciences, University of California, Los Angeles, California 99924, USA

T. S. Kemp, University Museum and Department of Zoology, University of Oxford, South Parks Road, Oxford OX1 3PS, UK

J. R. G. Turner, Department of Genetics, University of Leeds, Leeds LS2 9JT, UK

W. B. Provine, Professor of History and in the Division of Biological Sciences, Section of Ecology and Systematics, Cornell University, Ithaca, New York 14850, USA

K. S. Thomson, Yale University, Division of Vertebrate Zoology, Peabody Museum of Natural History, 170 Whitney Avenue, PO Box 6666, New Haven, Connecticut 06511-8161, USA

EDITORIAL

Evolution is how gene pools come to be what they are. Reflect upon it and you will find this to be a strictly sufficient definition, but so much the worse for strict sufficiency! It leaves so much unsaid. Evolutionary biology takes from, and gives to, all levels of biological science, molecular biology no less than ecology, as well as some other subjects not normally thought of as biological at all – economics for instance. Exciting as it is to embrace so much, it does not make for ease in keeping abreast. It is the aim of *Oxford Surveys in Evolutionary Biology*, through its annual volumes, to provide a medium whereby research workers and students are made aware of developments right across the wide field that constitutes modern evolutionary biology.

Volume Two of OSEB continues Volume One's tradition of encouraging thoughtful and reflective essays, rather than straightforward reviews of the research literature or reports of new research. Not that research is neglected. A. Grafen's geometric view of relatedness and W. S. Sutherland's discussion of measures of sexual selection both introduce new methods which field workers will want to use. H. A. Jerison reflects on some of the problems raised by research – of which he is, of course, the leading exponent – on the history of the vertebrate brain. T. S. Kemp analyses the logical difficulties inherent in any attempt to reconstruct the true branching pattern of phylogeny. G. C. Williams picks up a gauntlet thrown down in Volume One and shows that reductionism – which in some circles has become a nearly meaningless dirty word – is the key to the solution of some outstanding problems in evolutionary theory. This volume contains two historical essays on related themes. W. B. Provine examines the controversy between two of the three founding fathers of neo-Darwinism, Sewall Wright and Ronald Fisher. J. R. G. Turner takes a particular controversial issue in which Fisher was involved, that of the evolution of mimicry, and uses it to uncover some of Fisher's deepest convictions. Further essays on Ronald Fisher are planned for Volumes Three and Four. Finally, each volume will end with an extended essay book review on a collection of books on a theme. M. T. Ghiselin's review in Volume One of books on cladism is followed in the present volume by K. S. Thomson's thoughts stimulated by books on the relationship between evolution and embryology.

An activity of this kind, which is intended as a service to the scientific community, depends upon international cooperation. Our gratitude extends not only to our authors but to those who suggested fields and potential authors. We would like to encourage evolutionary biologists to continue to make such comments and suggestions to the editors or members of the editorial board.

RICHARD DAWKINS

A defense of reductionism in evolutionary biology

GEORGE C. WILLIAMS

Abstract

Reductionism is the seeking of explanations for complex systems entirely in what is known of their component parts and processes. I propose that reductionism is more likely to provide valid answers to questions about evolution, especially about adaptation, than recently suggested alternatives that invoke emergent properties at various levels of complexity. The currently most important reductionist device is to regard gene frequency change as the essence of evolution. Such change is a bookkeeping process that records past reproductive success and predicts what is likely to succeed in the future.

This record of reproductive success relates mainly to the reproductive success of individual organisms, but is theoretically capable of favouring adaptations at more inclusive levels. Also the independent gene pools of populations or taxonomic groups could have consistently different capabilities for long-term survival, so that the multiplication and extirpation of gene pools could keep a record of phylogenetic success. The general absence of any obvious functional organization in biological groupings beyond associations of close kin shows that such group selection is relatively weak as a factor in adaptive change. It could still be important in macroevolution, but a purely random proliferation and extinction of taxa may be all that is needed beyond evolutionary genetics to explain most of the major changes in the earth's biota.

The prevalent genes in a sexual population must be those that, as a mean condition through a large number of genotypes in a large number of situations, have had the most favourable phenotypic effects for their own replication. From this selfish-gene theory of adaptation comes a reductionist methodology known as the adaptationist programme. Its practitioners imagine that they understand a studied organism well enough to recognize certain of its features as components of some special problem-solving machinery. If other postulated components are not yet known, it is predicted that an appropriate investigation will reveal their existence. This is the most frequent kind of prediction now being practised in evolutionary biology and it has immensely enriched our understanding of organisms. The idea that a theory of evolution must predict future evolutionary change is unrealistic, but it has been the basis of some prominent criticisms of evolutionary reductionism.

A biological explanation should invoke no factors other than the laws of physical science, natural selection, and the contingencies of history. The idea that an organism has a complex history through which natural selection has been in constant operation imposes a special constraint on evolutionary theorizing. Every recognized evolutionary change had to be immediately useful to its possessor, every current feature had to be evolved from a slightly different earlier feature, and every part of the transition had to be adaptive. This is the only legitimate meaning that can be attached to the concept of phylogenetic constraint.

I also propose that only confusion can arise from the use of an animal-mind concept in any explanatory role in biological studies of behaviour. The absence of shared parameters (mass, distance, charge, etc.) between the mental and the physical must rule out any use of mental factors as causes in explanations of behaviour, and rule out any reductionist explanation for mental phenomena. For me, this is equivalent to the ruling out of scientific explanation.

Introduction

I argue here that various forms of reductionism, especially the selfish-gene view of natural selection urged by Dawkins (1976, 1982) can be richly fruitful for the understanding of adaptation and perhaps more useful in other evolutionary inquiries than such recent suggestions as the pluralism of Gould (1982 and elsewhere), the mentalism of Griffin (1981), or the hierarchical concepts of Eldredge and Salthe (1984). The selfish-gene theory of adaptation is, at least implicitly, a key component of biological thought, and of growing interest to philosophers and social scientists, although some grotesque misunderstandings of its concepts and terminology have found their way into print. Dawkins (1981) lists and rectifies several attributable to a single influential author. Some thoughtful criticisms (Wimsatt 1980; Sober and Lewontin 1982) are mainly based on what I regard as unrealistic expectations. Perhaps my efforts here will help to reduce the frequency of future misunderstandings.

And perhaps not. It may be a while before there is a real consensus on the value of various kinds of evidence and lines of resasoning, or even on the importance of various kinds of questions in evolutionary biology. Historians may one day marvel at the tardiness of the realization that life history attributes are subject to natural selection, and evolve no less than teeth and chromosomes, that the prevalence of meiosis presents a major theoretical challenge, or that many of the formulations of natural selection are applicable to cultural evolution. I attribute this tardiness to a persistent and widespread failure to make full use of the Mendelian formulation of natural selection. What is sometimes called modern Darwinism is a field in its infancy, at best.

Biological events: their genetic record, and the levels-of-selection controversy

Genetic differences between organisms may express themselves as differences in characters developed. These phenotypic differences may affect capability for survival and reproduction, and therefore the rates of transmission of different genes to later generations. This summary of the traditional theory of natural selection implies to some that the gene (that which is transmitted) is the ultimate entity of selection. Others, focusing on the different prospects for survival and reproduction, will object and claim that selection is ultimately of phenotypes. Still others might note that genes are identified with DNA molecules which, like organisms with phenotypes, merely reproduce at rates sufficient to balance destruction, whereas genealogies, breeding populations, and taxonomic groups may persist indefinitely. This may suggest a focus on persistent groups of organisms as the logically important evolutionary entities.

Such disputes can be purely semantic and arise without any disagreement on facts or logical connections between them, or they may reflect basic differences in perspective and on what would constitute productive research. I suggest that a formal separation of the informational (genetic) aspects of natural selection from its concrete ecological aspects (Table 1) may help in the avoidance of semantic distractions. I can then focus more clearly on real disagreements, such as those between myself and Wimsatt (1980), or Sober and Lewontin (1982). This approach (Table 1) emphasizes the distinction between *selection* and *response* to *selection* (Arnold and Wade, 1984) or Hull's (1980) *replication* and *interaction*. I think Hull's *interactor* more clearly descriptive than Dawkins' (1982) *vehicle*.

The strength and importance of the natural selection of various entities (such as those in the left column of the table) is one of the important issues currently argued in biology and I defend my own position mainly in the next section. The list is not complete, but can suffice to illustrate the distinction I wish to emphasize. It lists some of the levels at which important events may be happening, events that collectively constitute a history for each entity. The column to the right records my view on whether each sort of entity is a medium in which a record of such history is kept (whether it is suitable for the necessary bookkeeping).

Wimsatt (1980, p. 230) introduced the term 'bookkeeping' in a criticism of evolutionary reductionism in general and of my gene-centred reductionism

Table 1

Natural selection as a history of success and failure, and as a record of that history for various possible levels of selection.

		History	Bookkeeping
	gene	−	+
	genotype	+	sometimes
individual	phenotype	+	−
	family	+	−
	trait-group	+	−
	population	+	+ (gene pool)
	taxon	+	+ (gene pools)

4   George C. Williams

(Williams 1966) in particular. He spoke of my view as valid for a mere genetic bookkeeping and not for modelling future evolutionary change. I agree with Wimsatt's reasoning, but disagree as to its significance, because I think that discoveries, not future evolutionary events, are what we need to predict. The idea that bookkeeping has been taking place in the past is what gives the theory of natural selection its most important kind of predictive power, as I argue at length later on.

I have assumed a general agreement that at least a major part of the bookkeeping is done by shifting gene frequencies, so that the gene is very much a level of selection in the bookkeeping sense. I would assume (idiosyncratically, perhaps) that this changing gene frequency is not a part of the history recorded by changing gene frequency. The history of the writing of history is not history, but a kind of metahistory or historiography. Of course, a DNA molecule has a history, but molecules are very much a part of what is taking place on the left side of Table 1, and I would consider DNA a concrete part of an organism. The gene is not the molecule, but the information coded by the molecule.

This distinction may seem pedantic, but I think it will soon prove necessary as the idea of natural selection is extended into fields for which it was not originally intended. In the realm of cultural evolution, for example, a recognizable item of information (*meme*, in Dawkins' terminology), need not be confined to a single medium of expression. A proverb may be coded by molecular subtleties in someone's brain as a pattern of ink on paper or as irregularities in grooves on a plastic disc. The information may everywhere be the same, despite the diversity of media. It may be transcribed to and from many different media as it proliferates or dies out in cultural evolution. It just so happens that in genetics the archival medium is always DNA.

Someday, exobiologists may be able to analyse genetic systems other than those based on DNA, perhaps with one or more kinds of reverse transcription in a life cycle. Even in current terrestrial biology, the distinction between a gene and its medium may have some value. Michod (1983) objected to Dawkins' gene-as-replicator concept by pointing out that not all of the structure or information content of a DNA molecule is replicated. Its pH-sensitive tertiary structure is lost when the pH changes. Ghiselin (1981) objected to the idea that natural selection does not deal with genes, but with gene lineages. An abstract gene-as-information concept would not be vulnerable to such criticism.

In all essential features, a gene is an ideal entity for bookkeeping. It has an extreme stability and permanence by virtue of the precisely engineered replication processes of DNA. No message carved in Egyptian or any other stone can rival the antiquity or legibility of genic messages we carry in our chromosomes. The engineering of mitosis and meiosis ensures that every dividing cell transmits every gene to every daughter cell and that every gene has a 0.5 probability of going into each gamete. Each zygote then gets half its genes from its mother and half from its father. Thus the genes keep a precise record of the reproductive success of every individual.

No such precise bookkeeping need be found at the next level in the table, that of individual genotype. Genotypes get thoroughly shuffled, and therefore effectively erased, in sexual reproduction. Thus, if a population has no other way to reproduce, genotypes cannot be a unit of selection in the bookkeeping sense. Where asexual reproduction is possible, history can be recorded in the changing abundances of clones. Hence, my equivocation in Table 1.

Genotypes have dual significance as genetic environments in which a gene temporarily resides and as sets of instructions for producing phenotypes. A gene's phenotypic expression will differ in different genetic environments. Thus, its effects are thoroughly context-dependent, as Sober and Lewontin (1982) emphasized and illustrated with the simple one-locus example of heterozygote advantage. Once a stable numerical balance is reached by the interacting alleles the heterozygote will continue to transmit genes more effectively than either homozygote, but there is no change in gene frequency. History continues to be made, but it does not get recorded.

This example nicely illustrates the difference between the concrete and informational aspects of natural selection considered in the two columns of Table 1. It is an extreme example, dependent for its validity on an improbable degree of stability of the average fitnesses of three genotypes. Even if this condition is occasionally approximated, it in no way rules out the gene as a level of selection, in the bookkeeping sense, for other loci with other sorts of fitness distributions.

The phenotypes produced by different genotypes will vary in their fitness for survival and the transmission of genes. Thus, phenotypes are of central importance among the happenings of evolutionary history. They can play no role in the bookkeeping because, even in a clone, the successive generations of phenotypes may be markedly different because of environmental variables that affect development. Such instability is incompatible with the keeping of books. Banana trees reproduce clonally. The genotype of any banana may be almost exactly that of its ancestor of ten generations ago. Its phenotype (height, number of fruit, etc.) may be markedly different from that of its genetically identical parent.

It is true that, collectively, the bananas today may look much the same as those of many generations ago. There can also be a compelling familiarity in the faces that look out at us from centuries-old portraits of organisms that did no cloning. There seems to be what is often called a persistence of type or pattern. The term 'persistence' is misleading. There is a recurrence of type or pattern. The exact replication of immortal human genes results in the recurrence, generation after generation, of temporary human genotypes and phenotypes. The distinction is fundamental and only confusion can result from labelling as persistent or permanent that which merely recurs.

The next level to be considered (Table 1) is the family. As a recognizable entity, it never lasts more than one generation in most organisms, or more than a few in any. So it cannot be a unit of selection in the bookkeeping sense. It obviously can be in the sense of significant variation in survival and productivity. A common example (Wade 1979) would be pairs of monogamous birds breeding in a woodland, each with its own territory and its own nest. As Wimsatt (1980) pointed out, the natural selection taking place here may be mainly based on the survival and productivity of whole family groups. This reproductive structuring of the population would be ignored in simple models of gene transmission by the reproductive activities of individuals.

The kin selection models introduced by Hamilton (1964) and refined by numerous later workers provide a way of considering just this sort of situation as an example of selection at the individual level. I presume that kin selection and family selection models (Wade 1979) would often give the same answers. Which should be used may be partly a matter of taste and facility. I would prefer kin selection because it can deal with relationships both between and within families. If there is any possibility of cuckoldry there will be a lower average degree of kinship between a male bird and the young in his nest than there is for the female. This may result in the evolution of a lower degree of parental devotion for the male than for the female. It would also be expected to result in male adaptations against the possibility of cuckoldry. These phenomena would not be explained by any family-selection models.

This example illustrates one reason why it is misleading to include kin selection as a kind of group selection. It is not merely selection between family groups. It can operate with or without the formation of groups, and it can influence the evolution of interactions within any groups that do form. Even when groupings of relatives occur, the main usefulness of the theory of kin selection may relate to matters other than the relative survival and productivity of such groups. Consider what might be considered a tightly circumscribed group of kin, the seeds in a fruit. The seed embryos of many species can bear a variety of relationships to each other and to the fruiting plant, because they can be produced in a variety of ways: out-crossing, selfing and other kinds of inbreeding, and apomixis. The relationships between each such pair of individuals is a separate evolutionary problem, which could easily be overlooked by someone who merely considers the varying survival and productivity of fruit.

Even a single seed may inflict a nightmare of sociobiological problems. It is made up of parental sporophyte tissues in the seed coat, remnants of separate, but perhaps related gametophytes, the descendant sporophyte embryo, and a triploid endosperm. The effects of kin selection and other evolutionary processes on the reproductive cytology of seed plants is beginning to get some belated attention (Burnham and Stout 1983; Queller 1983; Willson and Burley 1983).

Trait groups (Wilson 1980) are temporary collections of individuals that interact in some ecologically important way. Being temporary they cannot participate in the bookkeeping, but as Wilson has clearly shown, they can certainly generate information that goes into the genetic book. All that is needed is some genetic difference between groups and for some combinations of genotypes to be more productive of offspring, either directly through higher rates of reproduction or indirectly through higher survival rates. Sibling associations are probably the most important kinds of trait groups and family selection the most prevalent kind of trait-group selection.

A population is a long-lasting collection of individuals, typically hundreds or more, which maintains itself for many generations by its members' reproduction. Populations can become extinct and they can also reproduce. A single widespread population can become several small ones by becoming discontinuous geographically. It may also reproduce by dispersing colonists to unoccupied areas. If some populations of a species are doing better than others at persistence and reproduction, and if such differences are caused in part by genetic differences, this selection at the population level must play a role in the evolution of a species.

Reading of the literature on variation among populations leads me to suppose that their genetic differences can persist over periods of time of evolutionary significance. So information generated by selection among populations need not be quickly erased. I conclude, after a sequence of negatives on the right side of Table 1, that there is bookkeeping going on here. The bookkeeping consists of the proliferation of some sets of gene-frequency values and the disappearance of others. A gene pool, no less than a gene, can potentially persist forever and is therefore a suitable medium for writing a record. Wimsatt (1980) gives a clear presentation of principles of heredity applicable to lineages of separate populations and of the adequacy of these principles for the operation of natural selection at the population level.

If selection can be effective at the level of populations in general, this must surely include those populations that have diverged enough to lead biologists to recognize taxonomic distinctions. This is the final level in Table 1 and what has recently been called species selection (Stanley 1975). I prefer the term taxon selection. Once separate gene pools are established, selection can operate between them whether the separation results from ethological incompatability or an intervening mountain range. It matters little what taxonomic rank is reached by the diverging lineages. Subspecies can replace each other, and so can orders and classes. The same graphic or mathematical models are applicable at all taxonomic levels.

For detailed discussion and partly contrary views on taxon selection see Maynard Smith (1983, p. 280–1) and Vrba (1983, 1984).

The importance of group selection

I have argued earlier that events at many levels, from genotype to taxonomic group, may be important for natural selection. I also argued that it is only at two levels, gene frequencies in populations and the different collections of gene frequencies in separate populations, that a record of such events is likely to be kept. If it is agreed that these two kinds of genetic bookkeeping are taking place, the next obvious question would be on their relative importance. The question is complicated because there can be more than one kind of importance. Here I will deal with two: importance for producing and maintaining adaptation, and importance for determining the properties of the earth's biota. My main interests have always been in the first kind and my approach to evolution through what Gould and Lewontin (1979) called the 'adaptationist programme'. I view an organism as a participant in a contest and it is as such that I wish most to understand it. Other biologists no doubt have other interests.

My assessment is that selection at any level above the family (group selection in a broad sense) is unimportant for the origin and maintenance of adaptation. I reach this conclusion by simple inspection. An organism appears to me to be trying to maximize its genetic contribution to future generations of the population to which it belongs. I see little to suggest that it is playing any subordinate role in the interests of that population. It participates in the game of life as an individual contestant, not as a member of a team.

Detail after detail of an organism seems designed for the well being and reproduction of that organism and no other. An understanding of this point was probably reached by the first thoughtful protohominid to contemplate its own body. Perhaps, as it examined its hand, it exclaimed something like:

> Come now, let us investigate this very important part of man's body, examining it to determine not simply whether it is suitable for an intelligent animal, but whether it is in every respect so constituted that it would not have been better if it had been made differently,

as Galen did hundreds of millennia later (May 1968, p. 72). Perhaps Galen's can stand as the earliest explicit formulation of the optimization principle in biology. The philosopher David Hume (1779) on contemplating functional relationships within more limited regions than the hand was led to exclaim, 'What a prodigious display of artifice'. Apparently, attitudes that underlie the adaptationist programme are not an aberration of the last few decades, as Gould and Lewontin (1979) claim. The intricacy and subtlety of the self-serving machinery of any organism shows that varying success at self-service readily registers itself in the genetic record and that this is an evolutionary process of great power.

No such organizing power for group selection can be seen at the level of populations and communities. Instead of the precise co-operation found among separate parts of a single body, we see that different members of a population mainly compete with each other. It may be possible to remove large numbers of individuals from a population, even with considerable bias for sex or age, without any obvious impairment resulting. The individuals not removed are more likely to show benefits from reduced competition. For comparison, try removing a large part of the body of a single organism.

Sometimes removal of a species from a community can have drastic consequences for those that remain. Removal of the dominant coral from a reef or of redwoods from a redwood forest will seriously disrupt associated communities. Such results usually follow only for species with gross physical effects on others. Removal of rocks from a rock garden would be little different. For most of the many species that inhabit a community without accounting for a major share of its biomass or metabolism, the extirpation of one will make little difference to most of the others. A community may provide many examples of mutualistic dependencies, but these do not, in the aggregate, provide a functional organization for the community as a whole. Compare the removal of a randomly selected species from a community with the removal of a randomly selected organ from an animal. Similar views of community ecology have expressed by Engelberg and Boyarsky (1979) and by Hoffman (1983), contrary ones by Patten and Odum (1981).

Functional subordination of component species is not normally found in natural communities, while other possible kinds of order, such as those expected of competitive exclusion, may be difficult to detect. Simberloff's (1982) survey of information on introduced species failed to find evidence of expected extinction among close competitors or of competition as a factor in the success of introduction. Extirpation of prey by introduced predators is more frequent. Yodzis (1982, 1984) showed that establishment of species in vacant habitats is mainly determined by what is introduced, with little subsequent effect from competition between species. Systems that operate normally when assembled of randomly-selected components in a random order may have the kind of statistical organization expected of randomness, but not the kind expected of natural selection.

The unimportance of group selection for adaptation need not imply that it is unimportant in other ways. As Leigh (1977) has pointed out and Wilson (1980) argued in detail, the total selfishness of every individual in a community need not mean that group selection is without effect in the evolution of that community. There are many different kinds of selfishness. If the kind found in species A is more destructive of long-term species survival than that found in B, B is more likely to persist as a member of the biota and to give rise to new species in the future.

Even if the natural selection of populations and taxa has only a thousandth part of the potential of individual selection for causing change, its effect in, for example, a million years could be appreciable, just as individual selection can produce appreciable effects in a thousand years. The biota after 10 million years could be a grossly biased sample of what it might have included if all species or a random sample of them had survived. The macroevolutionary effect of a selectivity in taxonomic persistence is now beginning to get the attention it deserves (Fowler and MacMahon 1982; Gould 1982; Van Valen 1975). It may be that the adaptationist programme based on selfish-gene reductionism will have little to contribute for an understanding of macroevolution.

Taxon-selection theories may serve no better. A serious difficulty for evaluating taxon selection as a force in macroevolution is that even random extinction, which I define as extinction of lineages with no consistently demonstrable differences between what survives and what dies out, can be a factor of great importance in macroevolution (Raup and Gould 1974). It is unlikely that more than a dozen of the vertebrate lineages that survived the Devonian have left any modern descendants. Were these survivors especially well favoured by features that might someday be shown to provide long-term advantages? Or were they just a random sample from the thousands of possibilities? Chance factors can be decisive with such small samples. If it had been a different sample dozen, the subsequent history of the vertebrates and all other groups would obviously have been radically different. Perhaps a chaotic sort of historical contingency is a more important macroevolutionary factor than group selection.

The less important kinds of prediction in evolutionary biology

I suspect that, to a layman and apparently to some philosophers and biologists, an obvious role for a theory of evolution is to predict evolution. This assumption underlies criticisms of selfish-gene reductionism levied by both Wimsatt (1980) and Sober and Lewontin (1982). Both have shown with detailed arguments, that such reductionism is incapable of predicting future gene frequencies at even a single locus. They are entirely correct in their reasoning.

Others do not regard this limitation as surprising or important. Wasserman (1981) argued that it is neither expected nor required that a theory of evolution predict evolution. Dunbar (1982) was more emphatic and referred to the prediction of evolution as a vacuous problem. Both recognized that a population is not a self-contained system in which some future state will follow deterministically from a current state. No biological theory could have predicted the global climatic changes and environmental disruptions of the Pleistocene. No current level of understanding can predict whether an important new mutation will occur and escape random loss or at what locus it will occur.

However, predictions of evolutionary change are routinely made, and they can play a key role in the planning of experimental investigations. This is obviously true for those who use experimental models of evolution. Wade (1979) predicted, with noteworthy success, evolutionary changes in experimental populations subjected to artificial group selection. Wade's and other experimental populations are carefully made to approximate self-contained systems in which internal cause-effect relations can be shown. Important insights can come from such work, but it should be realized that insights are not evidence and that evolution in an experimental population is only a model of evolution in nature. If an experimental population yields a result similar to that of a mathematical model, the result thereby becomes more robust, but not yet a conclusion based on evidence. The only experiments that can provide evidence on evolutionary history are those set up by nature in the remote past. The monitoring of these natural experiments is called the comparative method and is my focus of attention in the next section.

Prediction of evolutionary change, in the form of changing frequencies of Mendelian unit characters, is sometimes made for natural populations. I took part in one such study (Williams and Koehn 1978) based on the prediction that certain kinds of genetic change would occur independently at different loci in eel (*Anguilla*) populations and that these changes would be evident in comparing certain sorts of samples. We had no hope of gaining information on what eels would be like in the remote future, or even of identifying environmental factors responsible for the changes. We merely sought evidence on the population structure and intensity of natural selection acting on the eel, and we reasoned that the expected slight differences in gene frequencies would provide this evidence.

Applied biologists also make predictions of evolutionary change in economically important species. They predict that crop plants and livestock can be made to evolve resistance to diseases. They predict that insect pests will evolve resistence to insecticides, weeds to herbicides, and pathogens to antibiotics. Such changes are usually expected from prior experience rather than theoretical modelling. When the changes can be ascribed to shifts in frequencies of specific genes as a result of specific environmental stresses they can be cited as models of the microstructure of the Darwinian process, but not as evidence on its long-term effects.

These predictions of short-term genetic change in extant populations, whether experimental, agricultural, or natural, are the first of what I shall label the less important kinds of prediction in evolutionary biology. The second kind derives from theoretical models of phylogeny and predicts sets of characteristics not yet investigated. Implied predictions of this sort are as old as the idea of evolution. Any modern physical anthropologist has in mind a rather detailed picture of the latest common ancestor of man and chimpanzee, and most would be in rather detailed agreement on this picture. When we can examine well preserved fossils of this ancestor we can confirm or refute the consensus and resolve any minor disputes.

Unfortunately, such a decisive investigation is not likely to happen in the near future. It may be a long time before the required fossil is dug up. If and when it is found there will no doubt be some dispute about its identity as the most recent ape-human ancestor. Perhaps it already sits unrecognized in some museum. Galileo's enquiry into the mechanics of free-fall was easy compared to Darwin's into human origins. He had little doubt about his ability to drop weights from measured elevations on his tower and to time their falls. Darwin had no basis for optimism on his ability to find and identify fossils that could test his phylogenetic models. Historical and physical theories can be equally predictive, but the predictions grossly unequal in ease of testing.

Prediction of what the characteristics of a fossil ancestor will turn out to be, if and when it is possible to find out, is merely the most obvious kind of prediction from a proposed phylogeny. A detailed cladistic model for a large monophyletic group may include many statements and implications about times of origin of characteristics common to some, but not all, members of the group. Phylogenic theorists have recently developed objective methods for predicting the results of enquiry into characters of subgroups not yet examined, or into new sets of characters of those already well known in other respects (Eldredge 1981).

These two kinds of prediction are unimportant only in the limited sense of being made less often than the next kind to be discussed. Only a minority of evolutionary biologists devote their time to predictive cladistics and phylogenetic systematics. The other predictions, of short-term changes in actual populations, are perhaps more common, but are made for artificial systems or for reasons other than the testing of evolutionary explanations. Such predictions can, of course, play major roles in important investigations.

The more important predictions

The primary role of theory in evolutionary biology, as in any other science, is to predict the outcome of investigations not otherwise predictable. In this way, it helps investigators to direct their efforts into projects that are likely to be rewarding. It tells us where and how to dig for treasure, sometimes in a more than metaphorical sense, as in archaeology (Watson *et al.* 1971). The treasure might be some important new insight or theoretical refinement, a common result of the failure of prediction. Verified predictions provide further confidence in a system of beliefs as a basis for further work. Sometimes they lead to discoveries of great practical significance, as with Franklin's work on lightning or Redi's on the biogenesis of insects.

The important predictions in evolutionary biology derive from a concept of adaptation based on a selfish-gene model of natural selection and take this form: given that all organisms throughout their history have been subject to this kind of genetic bookkeeping, any given organism is expected to have certain properties and not have others. The properties are expected to form a closely-optimized strategy for the maximal proliferation of the genes that directed the development of the organism. No compromise with any other goal, like the survival of the species, is expected. Inspection of the organism will disclose whether it does or does not have the predicted properties.

A strategy can be said to be optimized only in relation to the rules of the game being played and to other necessary constraints. An important constraint arises from the gradualist view of adaptive change. Optima will never be more than local optima. If a species of eel has an average of 100 vertebrae and standard deviation of one vertebra, I would assume that, for the way of life of this species, 100 is better than 95 or 105. I would not be at all confident of its being better than 80 or 120. Perhaps one or the other of these numbers would facilitate certain kinds of movement of great value in burrowing, predator avoidance, or some other task.

Galen seems not to have appreciated this constraint. I am inclined to edit the passage quoted above to end

> . . . it would not have been better had it been made different by a few standard deviations in any feature.

A more recent neglect of the same constraint is Rothstein's (1982). He argued that birds that raise such nest parasites as cowbirds violate the optimization principle. They would be better adapted if they behaved differently and evicted the parasites. This statement about their behaviour is true, but not evidence against optimization in its legitimate relationship to a local optimum. The cowbird-rearing behaviour of a flycatcher violates local optimization only if a slightly different average behaviour, one within reach of the variation commonly found in the population, would enable it to raise its own young more often.

No such conclusion is supported by Rothstein's observations. The important conclusion that they do support is the validity of gradualism. A more adaptive form of behaviour is possible: that shown by other species that normally reject cowbird eggs. Unfortunately, there may be no way for the flycatcher to reach that superior condition by the Darwinian process of accumulating slight changes, each of which must be beneficial to its possessor. To deny this tenet of gradualism is to deny phylogenetic constraint. If populations can commonly take big steps in evolution we may expect them to regularly step from lower to higher adaptive peaks. Saltationist views of evolution would indeed provide the *Panglossian paradigm* deplored by Gould and Lewontin (1979).

Most evolutionary biologists seem to recognize that optimization is valid only in relation to local optima. Many no doubt realize that they can refute Galen's dictum and design a better hand, but that this sort of exercise has no bearing on biological adaptation. The prevalence of functionally arbitrary or maladaptive features of organisms was historically important for the acceptance of natural selection in place of intelligent design. Any vertebrate would be improved by having entirely separate feeding and respiratory systems to prevent it from choking on food, but this is not a local optimum for any vertebrate. Low-fecundity vertebrates might be better off adopting exclusively asexual reproduction so as to avoid the cost of meiosis (Williams 1975), but small steps toward that condition would be maladaptive. Slaves in an ant nest might be better adapted if they abandoned their masters and went back to their own queens, but such rebellion is not a local optimum that could be produced by adaptive modification of present behaviour (Gladstone 1981).

It is especially clear for contests with other organisms that strategies favoured by selection may be quite different from ideal adaptations that we might imagine. No doubt it would be less adaptive for a pregnant mammal to abort than to give birth at full term and raise her litter to independence. This ideal may be unavailable in an environment that includes a male for whom it is adaptive to kill the litter as soon as it is born, and abortion may be the favoured strategy (Labov 1981). Male-female and other kinds of conflict produce winners and losers, or else a compromise optimal for neither party. Conflict is better analysed initially by game theory than optimization (Maynard Smith 1982). Identification of females as losers and of abortion as their stable strategy would then lead to predictions based on the assumption of optimization for stimulus thresholds, timing, and other parameters of abortion.

Like other ultimately necessary factors, the phylogenetic pathway to an adaptation can be neglected in some preliminary approaches to its understanding. This justifies (temporarily) Maynard Smith's (1977) use of game theory to analyse mate desertion options for each sex in relation to reproductive processes in the opposite sex. He went on to consider why monogamous mammals never have male lactation, which would be adaptive according to his analysis. The answer I suggest is that intermediate stages between no male lactation and optimal male lactation have not yet been favoured in any mammal. No male mammal would have an advantage from slightly better developed mammary tissue, or from being slightly more likely to carry his young in a position that would facilitate nursing. In other words, adaptive male lactation is not a sufficiently local optimum. Enquiry into whether an adaptation is likely to evolve must ultimately deal with the question of what it can evolve from in an ancestor that lacks it.

Ridley (1978) took the necessary step of relating optimization to phylogeny in his comprehensive study of the evolution of male care of offspring. In the case of fishes, he found that egg guarding by males readily evolves in groups in which males are territorial, a common phenomenon. Gittleman (1981), in a quantitative study of phylogeny of fishes, argued that guarding by females alone evolves mainly from guarding by both sexes, a rather infrequent derivative of guarding by males. This partly explains the more common occurrence of paternal than of maternal care in fishes, even though maternal care would often be optimal, according to Maynard Smith's (1977) reasoning.

In using the optimization concept for understanding adaptation it is necessary to distinguish 'strategies', which organisms try to optimize, from 'winnings', which they try to maximize. The characteristic longevity, for instance, would not usually be a characteristic subject to optimization, but rather the variable result of organisms' attempts to avoid death as long as possible. Body size for many organisms may also be a highly variable measure of winnings, like longevity. Success at reproduction is often size dependent, a common example being a simple relationship between female fecundity and body mass. Optimum size in a population may well be one that is seldom attained and the average adult may be much smaller than that optimum. In such a population we expect optimization, not of mean size, but of mechanisms of resource capture and allocation that permit rapid growth. If a larger size evolves, an adaptive rescaling of structure might permit a larger optimum, which would remain well above the mean. The optimum level of out-crossing for mate choice may likewise change as the mean genetic distance between mates changes and remain much greater than the mean (Waddington 1983).

There is no simple rule for distinguishing strategies from winnings. In some groups body size may be tightly constrained for functional reasons and rather precisely optimized. In birds, for example, adult size may have little variability in a population, and the mean and optimum may well be nearly the same. Developing organisms often outgrow their important predators, so that size would be a measure of 'winnings', but sometimes the opposite occurs. Planktonic crustacea in freshwater habitats in which the main predators are fish of more than a few grams may be largely free from attack if they mature and complete their life cycles below a certain size (Brooks and Dodson 1965). Adult body size may represent an optimum compromise between conflicting values of minimizing predation and maximizing fecundity. The best way to distinguish optimized from maximized characteristics is to gain a detailed familiarity with the natural history of the organism under study. This difficulty is further evidence of the immaturity of the study of adaptation.

On the basis of such familiarity a biologist may come to appreciate the problems that an organism must overcome to achieve its ultimate goal of maximal proliferation of its own genes. Known features of the organism may look like components of strategies for solving some of the problems. The study passes from the realm of natural history to that of science when the investigator postulates an undocumented remainder of a strategy. The next step in the scientific enterprise is to look and see if one or more of the postulated features are actually there. Thus, from a proposed strategy of selfish-gene proliferation can come predictions of an organism's properties that would not otherwise have been suspected. That this is the usual predictive role of the theory of natural selection in evolutionary biology is now widely appreciated (Brown 1982; Clutton-Brock and Harvey 1979; Mayr 1983; Thornhill 1984).

In this way Reichman and Aitchison (1981) predicted angle of ascent of mammals on mountain sides as a function of slope and body mass; Rhoades and Bergdahl (1981) predicted the pattern of distribution of toxic substances in nectar as a function of community diversity and other factors; Vitt (1981) predicted a relationship between clutch size and refuge behaviour in lizards; and Bertness (1981) predicted rates of somatic v. gonadal growth in hermit crabs as a function of the size of the shells they inhabit. These are merely four of many examples that could be cited from a single biological journal in a single year. The biological literature must now contain thousands of such attempts to understand adaptation via a thoroughly reductionist theory of natural selection, and to make and test deductions from the proposed understanding. This is overwhelmingly the most important kind of prediction now being practised in evolutionary biology.

This kind of quest for understanding has recently been criticized by Gould (1980), and Gould and Lewontin (1979), who claim that it is not possible to specify problems faced by organisms and then to identify adaptations by which they solve them. A somewhat different view was expressed by Popper (1974, p. 272) who claimed that '. . . we can describe life, if we like, as problem solving and living organisms as the only problem solving complexes in the universe'. I believe that it is already clear that attempts to understand organisms as problem-solving complexes with the aid of a reductionist view of the origin and maintenance of the problem-solving machinery have been richly rewarding, despite the pessimism expressed by Gould and a few others.

Just So Stories

If Rhoades and Bergdahl (1981) had called their article 'How the ragwort nectar got its alkaloids', it would have been pointedly reminiscent of many of Rudyard Kipling's classic children's tales entitled *Just So Stories*. Gould (1980) called attention to the resemblances of modern adaptation studies to Kipling's works with the implication that this impugns the seriousness and importance of such studies as Rhoades and Bergdahl's. I would propose the opposite, that the resemblances should suggest to students of adaptation that they might find some lessons in Kipling. In the *Just So Stories* there are many examples of explanations for special features of animals. From these explanations predictions might be deduced and subjected to empirical test. Kipling's proposals differ from those of modern Darwinism in being based on a system of genetics and ecology quite different from selfish-gene reductionism. This difference affects the nature of the explanations generated, but does not affect their status as systems from which hypotheses can be deduced and tested.

For example, consider Kipling's story 'The Elephant's Child'. It stated that remote ancestors of modern elephants were much like those of today except that the nose, while large and fleshy, was not especially elongate. One day a young elephant, from which all modern ones descend, lowered its head for an intimate conversation with a crocodile. Instead of conversing the crocodile seized it by the end of the nose. The elephant's efforts to pull loose stretched its nose into the shape we find in modern elephants. In Kipling's system of inheritance this trauma was transcribed into the germ plasm so that all modern elephants show its effects.

So, according to Kipling the shape and internal structure of the elephant's trunk is that expected from the plastic stretching of a more compact structure. Either it does or does not have these features, an issue that observation can decide. As an alternative, we could use the premise that the trunk is functionally optimized for one or more kinds of usefulness in the proliferation of the genes that direct its development (prehension, for example). I have some appreciation of the complexity of the musculature of this region and little doubt about which theory a dissection of the elephant's trunk would support.

However, suppose it turned out the opposite to my expected verification of the functional premise. Suppose the shape and inner structure of this striking feature proved to conform in many details to that expected of stretching an ordinary mammalian nose and made no sense functionally. Suppose such trunk uses as lifting logs and squirting water are accomplished by an accidental usefulness, much as a chair can be used as a door stop or a coat rack. We would have to conclude that Kipling's is the preferred theory. We might then look into the matter further, as I would expect of users of the adaptationist programme, and discover that the end of the trunk bears scars with just the pattern we would expect from a crocodile bite. Kipling's theory would be attractive indeed.

The obvious reason why no biologist is busy testing Kipling's *Just So Stories* is that they are based on fanciful and discredited assumptions about heredity and development. The mundane and abundantly documented generalizations from which Darwin deduced his theory of natural selection (offspring may be produced at more than replacement rates, like begets like, etc.) are much more acceptable. A less obvious, but likewise important reason, is that natural selection is, by its nature, a force tending towards functional improvement. It offers direct aid in dealing with Hume's 'prodigious display of artifice'.

The methodological identity of evolutionary biology and other sciences

Shapiro (1981) followed the adaptationist programme in predicting that a butterfly species would lay its eggs at certain times and in a widely dispersed pattern. On testing these predictions he found something quite different, clumped distributions and some apparently maladaptive timings of events in the life cycle. As a result, he took a course of action condemned by critics of the programme. He consulted his imagination and found an *ad hoc* assumption that would excuse the discrepancy. He proposed that the weedy fields in which he studied the species are a recent human artifact and an abnormal environment, but that the observed seasonal schedule and egg-laying patterns would be adaptive in a hypothesized ancestral environment. He does not seem to have considered that some basic premise of his theory of adaptation might be wrong.

What would a well-disciplined physical scientist do in a comparable situation? Suppose an engine, designed according to accepted physical principles, is found not to work in the expected manner. The engine's failure might then be cited as evidence of a basic flaw in physics. I think it more likely that an engineer would first look to see if the engine had a wet distributor or an empty fuel tank. If he fails to confirm suspicions of this sort he might postulate that a factory worker misplaced some of his solder or a designer one of his decimal points. Any sane engineer would first look for trivial rather than profound causes of failure, and test them in the expectation of confirming at least one. In his use of a plausible *ad hoc* assumption to explain the discrepancy between observed and expected egg-laying behaviour, Shapiro provided not only a possibly important idea, but also evidence of sanity.

Failure of a prediction may ultimately provide the underlying theory with a greater triumph than would result from confirmation. Early in the nineteenth century, when Newtonian mechanism had not yet reached the pinnacle of esteem that it later achieved, astronomers tried using it to make really precise predictions of the future positions of planets. A serious contradiction was soon noted between observed and expected positions of Uranus. A possible reaction was to propose that one or more of Newton's laws was wrong. Perhaps the inverse square law broke down for objects as far from each other as the Sun and Uranus. Another possibility was to dream up an *ad hoc* assumption that would explain the discrepancy and let people go on believing in the simple truth of Newtonian mechanics.

At least two people, Adams in England and Leverier in France, did just that (Krauskopf 1953). They proposed that it was wrong to assume that the only important gravitational forces acting on Uranus were from the sun and known major planets. They imagined another force of a certain magnitude and direction, from a hitherto unknown planet, which might explain the discrepancy. With this *ad hoc* addition to their previous list of axioms they made a new prediction, that a certain kind of scrutiny of a limited region of the sky would disclose the source of the imagined gravitational force. A test of this prediction led to the discovery of Neptune.

Note that this classic example of an astronomical prediction did not predict a future astronomical phenomenon in a way that would meet a demand like Sober and Lewontin's (1982) for a theory to predict future gene frequencies. Neptune was there all the time, and sooner or later would have been discovered by accident, as Uranus had been a century earlier. The prediction was that a certain kind of investigation would have a specified result of great interest.

Less spectacular, but analogous, episodes are common results of *ad hoc* assumptions in the adaptationist programme. Spontaneous and unreciprocated altruism towards non-relatives is impossible according to selfish-gene expectations. Thus, the observation of certain male fishes tending eggs fertilized by other males was contrary to expectation. The theory could be saved by the gratuitous assumption that the females prefer to mate with males that already have eggs in their nests. This previously unsuspected phenomenon has now been verified for several species (Ridley and Rechten 1981; Downhower and Brown 1981; Lawrence Unger, personal communication). This theory-based discovery by Downhower and Brown for the mottled sculpin led them to yet another *Just So Story* that might be called 'How the female sculpin got her mate-choice inclinations'. An imagined component of this story is that males are more likely to abandon single than multiple clutches. Tests verified this previously unknown aspect of sculpin behaviour and suggested a *Just So Story* on how the male sculpin got his clutch-abandonment behaviour, and so on.

Neither this theorizing about sculpins nor Shapiro's (1981) resort to the abnormal-environment excuse is likely to lead to the discovery of anything as newsworthy as a planet. However, it could well stimulate more predictions, which could in turn lead to important discoveries, and Shapiro's report states an intent to exploit this possibility. If he is right, the populations studied must now be subject to strong directional selection on oviposition behaviour, and a check on this inference could be rewarding. An identification of the ancestral environment as that in which the current egg distribution would be adaptive could be checked against independent evidence on the recent history of the species.

Criticism of students of adaptation who resort to *ad hoc* excuses for the failures of the predictions must arise from a mistaken view that predictions are tested to check on the truth of general theory. I suspect that this is almost never correct. The predictions are tested to check on the truth of an understanding of the phenomena investigated. This understanding always includes special assumptions in addition to those of the major theory, and it is these that are confirmed or modified according to the outcome of an investigation. In discussing thermodynamics, Tribus (1966) said '. . . within the framework of a theory it is not possible to disprove its principles. There are too many escape hatches'.

Maynard Smith (1978) recognized that no general theory of adaptation was being tested in such investigations as I have been discussing. He proposed that evidence against the theory could come only from direct experimental invalidation of its premises. An example might be the demonstration that much of inheritance is Lamarckian. The theory of natural selection could perhaps be falsified in this manner, but I think that another kind of downfall would be more in keeping with analogous historical developments. The Daltonian atom went out of favour, but not because anyone was able to show that any of its basic statements were wrong. No statement about an atom could be directly investigated. Dalton's atom was abandoned simply because the rather different Bohr atom was a much more useful explanatory device. As was pointed out long ago by Conant (1951, p. 48) in a history of the concept of the atom, '. . . a theory is only overthrown by a better theory, never merely by contradictory facts'. The adaptationist programme now so widely followed will be abandoned if and when we learn of a better device for coping with such phenomena as are seen in the interactions of male and female sculpins, or between various nectar-producing plants and their potential pollinators and thieves.

Other necessary kinds of reductionism

Darwinism was radically reductionistic from the start in maintaining that the biological world had been produced by a strictly material evolution. Even more extreme was its insistence that this evolution required only the most mundane processes for its operation. It is these processes, such as the prevalence of genetic variation and all organisms' potential for increasing in numbers, that are usually cited as the logical premises of Darwin's theory. This use of empirical generalizations from natural history as basic theoretical postulates would seem to make this theory different from theories in physical science, which are all based on proposals that originate in the imagination (atoms, ethers, idealized geometrical figures, etc.).

Empirical generalizations were indeed essential in Darwin's argument, but so was another premise that was not directly testable. It was essential for Darwin that only the mundane processes could ever be operative and that major changes could all be explained by extrapolating the mundane. This could only be reasonable with the newly acceptable geological conclusion that there was an enormity of time for the operation of the mundane processes. By his implied ruling out not only of the supernatural, but even of the unusual, Darwin made his theory theoretical, subject only to indirect test, and therefore like other scientific theories.

Darwin's list of mundane factors has been purged of a few items, such as inherited effects of use and disuse. Many other factors have been proposed for addition to the list over the last 125 years, some a thinly-veiled mysticism with erudite names, like aristogenesis and the Gaia hypothesis, others not so readily evaluated, like various of Gould and Lewontin's (1979) proposed additions or alternatives to the adaptationist programme. I accept with enthusiasm that for anything to evolve it must do so by modifying that which came before, and I heartily agree with Darwin's insistence that every stage in the evolution of any useful feature has to be useful to its possessors, not merely of potential use later. If the effects of bauplane or of phylogenetic constraint (Gould and Lewontin 1979) means more than this, the meanings escape me.

Some workers seem to assume that if structural variation in a taxon can be described algebraically, with some terms widely variable and others constant, the constancies must represent quantities that natural selection is unable to change (Gould 1984; Holder 1983; Huxley 1932). Holder is explicit in his belief that adaptive diversification would produce complex patterns '. . . at the whim of external environmental influences' and that mathematical uniformities in variability must be attributed to developmental constraints that make certain kinds of change impossible. I wonder how one distinguishes taxon-wide developmental constraints from taxon-wide consistency of function. An analogy to Holder's polar co-ordination model of vertebrate limb variation might be the use of geometrical parameters for describing variation in the wheels on vehicles. Wheels would be found to be immensely variable in some parameters, but rigidly constant in symmetry and eccentricity. Does this mean that factories must be incapable of producing tyres with different tread width at different places on the perimeter, or any departures from nearly perfect circularity?

I once insisted that '. . . the laws of physical science plus natural selection can furnish a complete explanation for any biological phenomenon' (Williams 1966, pp. 6–7). I wish now that I had taken a less extreme view and merely identified natural selection as the only theory that a biologist needs in addition to those of the physical scientist. Both the biologist and the physical scientist need to reckon with historical legacies to explain any real-world phenomenon. No matter how precisely the mechanical laws may predict the future motion of some planet, they do not suffice to explain how the planet got its present velocity, mass, and other measured attributes. A historical theory of the origin and evolution of the solar system is needed to complete the explanation for the planet.

The needs are the same in the microscopic world. Physicochemical laws explain much of what happens on a photographic plate exposed to radioactive material. They do not explain how the plate came to be exposed in the first place, nor can they, in principle, explain certain subatomic phenomena, such as why one atom rather than another emitted radiation. Both the positioning of the plate and radioactive substance, and the particular pattern of spots developed on the plate are historical contingencies beyond the reach of physics and chemistry.

History is an even more obvious element in the biological world. Unique events in the early history of life must have profoundly affected all subsequent evolution. No biological theory could have predicted the rise of the Isthmus of Panama, and consequent division of many continuous marine populations into separate Atlantic and Pacific moieties, each then free to evolve independently of the other. Chance events are important in evolution throughout the spectrum, from subatomic to planetary levels. Equally fit organisms may have unequal reproductive success. Genetic drift depends on such chance variation and also on such processes as Brownian movement, which must play a role in deciding which sperm fertilizes an egg or which chromosome goes into the egg rather than a polar body. The stochastic nature of the mutation process is universally recognized.

Mutation and drift are examples of historical contingency at the individual and lower levels, but once they happen they become, like the Isthmus of Panama, historical causes that influence subsequent evolution. An analogy is seen in the tendency for special genetic stocks of *Drosophila* to show less extreme phenotypes when maintained for many generations. Homozygosity for an abnormal allele at one locus causes favourable selection, at other loci, of genes that can compensate for the abnormality (Wallace 1985). I suggest that the term evolution should be used only in this sense, of a process determined by chance events plus the persistent biases of natural selection. To the extent that chance events lose their causal influence by being corrected by some control mechanism, the process should be considered developmental rather than evolutionary.

Individual ontogeny is canalized by control mechanisms that persistently correct perturbations from a prescribed course of development. To a lesser extent ecological succession is canalized and therefore predictable because of density-dependent feedback in the regulation of populations, and because of cumulative changes caused by each species' activities. I see no evidence that evolution is subject to such control, but others often do. Ever since the fixity of species was first questioned there have been biologists who have regarded the history of change in the earth's biota as, to a varying extent, developmental rather than evolutionary. Some examples are Lamarck (Elliott 1963), Teilhard de Chardirn (1959), and perhaps Huxley (1954).

There is no biological generalization so general that it can serve as a principle of the same rank as natural selection. The most fundamental features of cellular organization must be explained as characters evolved by natural selection, just like any other characters (Cosmides and Tooby 1981; Eberhard 1980). Mendel's laws of genetics can be used routinely as evolutionary premises only for organisms known to reproduce sexually and to have Mendelian heredity. Ultimately, it will be necessary to explain why reproduction is so often sexual and inheritance Mendelian (Bell 1982).

One final complaint about current biological thought. It is commonly assumed that the demonstration of a certain level of capability for information processing shows the presence of a mind. This assumption underlies much of the discussion of the animal mind by Griffin (1981) and Dennett (1983). Mind may be self-evident to most people, but I see only a remote possibility of its being made logically or empirically evident. As Watson (1982) pointed out in a review of Griffin's work, 'the author knows his thoughts by direct introspection. He knows other minds indirectly by analogy'. The animal-mind problem is merely a special case of the other-mind problem that has troubled philosophy ever since it began. I feel intuitively that my daughter's horse has a mind. I am even more convinced that my daughter has. Neither conclusion is supported by reason or evidence.

The remote possibility that I would concede, for producing real evidence for a mind other than my own, might take some such form as this: an investigator understands the nervous system (or other behaviour-controlling machinery) of a lover, or pet dog, or computer, or captive extra-terrestrial visitor. The understanding ought, on the basis of well established principles of physical science, to make the subject's behaviour highly predictable. Investigation reveals persistent discrepancies between prediction and observation. Further study shows the discrepancies to result from capriciously-distributed violations of some physical principle, perhaps the conservation of energy. Finally, it is shown that the physically capricious distribution of anomalies bears an orderly relation to statements that the subject offers as descriptive of its own states of mind.

This would be evidence that a mental realm exists beyond the physical, but capable of interfering with the physical. There would be no such evidence in a mere parallelism between physically lawful brain function and statements from the brain's owner. The statements would merely be one kind of behaviour that the understanding of the physical machinery could predict. Only if it violates physical laws would mind be a factor that biologists would have to deal with and could potentially deal with. To understand the behaviour of an animal they would need to know not only its physical attributes, but also a separate kind of data on mental attributes. Griffin's quest for 'a possible window on the minds of animals' would be an urgent one.

There is no such evidence for mind as an entity that interfers with physical processes, and therefore there can be no physical or biological science of mind. As Lewontin pointed out in his commentary following Dennett (1983), no kind of material reductionism can approach any mental phenomenon. Reductionism between all levels from the cosmic to the subatomic depend on shared parameters. Both a galaxy and an electron have mass, charge, velocity, and other properties in common. There is no such commonality between the physical and the mental. The power of positive thinking has never been measured in calories per second, nor a burden of grief in grams.

If the mental realm could ever be shown to interfere with the physical, such measurements would be possible and the two realms would have shared parameters. A big thought might create more calories than a small one, and thought would become an item for physical measurement and biological theorizing. Lewontin's solution to the non-objectivity of mind is to abandon reductionism. Mine is to exclude mind from all biological discussion. Both positions have no doubt been taken by various philosophers in the last few centuries. Mine was defended by Tinbergen (1963) and in 1891 by T. H. Huxley (Huxley 1902, p. 299).

Any vernacular terms used for biological phenomena must be considered shorn of whatever mentalistic implications they may have in ordinary discourse. If I say that an animal is hungry, this means that it is likely to eat if given a chance, and may go to considerable effort to find and secure something to eat. If I say that it is more friendly to its relatives than to other conspecifics, I mean that it processes perceived information about these other individuals in a certain way and that the outcome determines how friendly it will behave towards them. In a biological argument I would never use such statements to imply any view on what the animal might be experiencing in its own mind.

The whimsical recognition of mentalism is a frequent obfuscation in discussions of adaptation and especially of animal behaviour. In teaching such topics as kin selection, mate choice, and life-history evolution to undergraduates, I find it convenient to begin with botanical examples. Few students are inclined to suppose that an ovule self-consciously desires fertilization by one pollen nucleus rather than another, that a plant really loves its seeds, or that a seed desires a certain combination of conditions for germination. This experience then makes it easier for students to discuss analogous zoological examples as biology rather than soap opera. The same should be true for those philosophers (e.g., Dennett 1983) who think that a mentalistic concept of intention is necessary to explain the adaptiveness of animal behaviour.

An ingrained mentalism must underlie many of the more egregious misunderstandings of a selfish-gene reductionism, for instance Midgley's (1979). Unfortunately, Dawkins (1981) may have slipped into some mentalism in responding to Midgley. Instead of simply and decisively divesting the word calculate of implied mentalism, he said that he does not really believe that animals calculate. Surely any word that we use to describe information processing by a computer can be used to describe analogous functions in animals. If an entity confronted with a quantitative problem produces an accurate answer, surely it must have indulged in calculation. An animal need not express the answer symbolically any more than a computer need print a calculated result. It may make more functional sense for the result to go directly to the animal's muscles or to machinery controlled by the computer.

A horse named Clever Hans once excited great wonder by his apparent ability to do arithmetical problems presented either orally or in print. He expressed his answers either by picking the correct one among printed alternatives or by stomping the right number of times with a hoof. It was finally shown that Clever Hans solved problems by responding to unconscious cues from the questioner and his mathematical capabilities were considered disproved (Gould 1982). I prefer to say that he was unable to use Arabic numerals or German for any sort of analysis or communication, but that there is no doubt about ability to calculate. Every time he trotted across a field he rapidly performed a series of calculations far more subtle than long division.

Perhaps another example can make clearer the mathematical nature of animals' motor control. We would not hesitate to speak of calculations performed by a computer that controlled the aiming and firing of a cannon by processing information on a target's horizontal distance, altitude, and compass direction. This is exactly what is done by the brain of an archer fish when it spits with impressive accuracy through the water surface at insects on overhead vegetation. It need not worry about a Coriolis force that would affect the accuracy of long-distance cannon fire, but it needs a trigonometric function of the refractive indices of water and air in evaluating the apparent elevation of the insect.

For me the function of any biological theory is to help me understand flesh-and-blood (or cellulose) organisms and I find the inclusive fitness concept, directly derivable from the reductionist, selfish-gene theory of natural selection, a useful device. The concept may indeed need some refinements, as is painfully clear from Dawkins' (1982) criticisms. I have suggested some in this essay, such as the distinction between strategies and winnings. In this I depart from Dawkins, who would restrict the concept of fitness to the gene and not try to salvage its applicability to their interactors.

The history of science shows the inevitability of widespread emotional dissatisfaction with reductionist interpretations of the world. I suppose there are people who think that the theory of sexual selection robs the grossbeak's song of its music or that kin selection makes a mockery of brotherly love. There used to be those who thought that Newtonian optics destroyed the beauty of the rainbow (Medawar 1974).

ACKNOWLEDGEMENTS

Contribution number 537 from the Program in Ecology and Evolution, State University of New York, Stony Brook, New York 11794. This paper began as a presentation to a symposium on *Persistent Questions in Evolutionary Biology* at the University of Chicago in March, 1982. At that time I was helped immensely by conversations with William C. Wimsatt. Much improvement of a later manuscript was made possible by criticisms from Douglas J. Futuyma, David L. Hull, Richard C. Lewontin, and Elizabeth S. Vrba. My wife, Doris Calhoun Williams, helped in many ways with bibliographic problems.

References

Arnold, S. J. and Wade, M. J. (1984). On the measurement of natural and sexual selection: theory. *Evolution*, **38**, 709–19.
Bell, G. (1982). *The Masterpiece of Nature. The Evolution and Genetics of Sexuality*. University of California Press, Berkeley.
Bertness, M. D. (1981). Pattern and plasticity in tropical hermit crab growth and reproduction. *Am. Nat.* **117**, 754–73.
Brooks, J. L. and Dodson, S. I. (1965). Predation, body size, and composition of plankton. *Science*, **150**, 28–35.
Brown, J. L. (1982). The adaptationist program. *Science*, **217**, 884–6.
Burnham, C. R. and Stout, J. T. (1983). Linkage and spore abortion in chromosomal interchanges in *Datura stramonium* L. Megaspore competition? *Am. Nat.* **121**, 385–94.
Clutton-Brock, T. H. and Harvey, P. H. (1979). Comparison and adaptation. *Proc. Roy. Soc. Lond.* **B205**, 547–65.
Conant, J. R. (1951). *On Understanding Science*. The New American Library (Mentor), New York.
Cosmides, L. M. and Tooby, J. (1981). Ctyoplasmic inheritance and intragenomic conflict. *J. theoret. Biol.* **89**, 83–129.
Dawkins, R. (1976). *The Selfish Gene*. Oxford University Press, Oxford.
—— (1981). In defence of selfish genes. *Philosophy*, **56**, 556–73.
—— (1982). *The Extended Phenotype*. W. H. Freeman & Co., Oxford.

Dennett, D. (1983). Intentional systems in cognitive ethology: The 'Panglossian paradigm' defended (with peer commentaries). *Behav. Brain Sci.* **6**, 343–90.

Downhower, J. F. and Brown, L. (1981). The timing of reproduction and its behavioural consequences for mottled sculpins, *Cottus bairdi*. In *Natural Selection and Social Behaviour* (eds R. D. Alexander and D. W. Tinkle), pp. 78–95. Chiron Press, New York.

Dunbar, R. I. M. (1982). Adaptation, fitness, and the evolutionary tautology. In *Current Problems in Sociobiology* (Edited by King's College Sociobiology Group), pp. 9–28. Cambridge University Press, Cambridge.

Eberhard, W. G. (1980). Evolutionary consequences of intracellular organelle competition. *Q. Rev. Biol.* **55**, 231–49.

Eldredge, N. (1981). Evolution and prediction. *Science,* **212**, 737.

—— and Salthe, S. N. (1984). Hierarchy and evolution. *Oxford Surv. Evolut. Biol.* **2**, 184–208.

Elliott, H. (translator) (1963). *Zoological Philosophy* (of J. B. Lamarck). Hafner Publ. Co., NY.

Engelberg, J. and Boyarsky, L. L. (1979). The noncybernetic nature of ecosystems. *Am. Nat.* **114**, 317–24.

Fowler, C. W. and MacMahon, J. A. (1982). Selective extinction and speciation: their influence on the structure and functioning of communities and ecosystems. *Am. Nat.* **119**, 480–98.

Ghiselin, M. T. (1981). Categories, life, and thinking. *Behav. Brain Sci.* **4**, 269–83.

Gittleman, J. L. (1981). The phylogeny of parental care in fishes. *Anim. Behav.* **292**, 936–41.

Gladstone, D. E. (1981). Why there are no ant slave rebellions. *Am. Nat.* **117**, 779–81.

Gould, J. (1982). *Ethology*. W. W. Norton & Co., New York.

Gould, S. J. (1980). Sociobiology and the theory of natural selection. *Am. Ass. Adv. Sci. Symp.* **35**, 257–69.

—— (1982). Darwinism and the expansion of evolutionary theory. *Science,* **216**, 380–7.

—— (1984). Covariance sets and ordered geographic variation in *Cerion* from Aruba, Bonaire, and Curacao: a way of studying non-adaptation. *Syst. Zool.* **33**, 217–37.

—— and Lewontin, R. C. (1979). The spandrels of San Marco and the Panglossian paradigm: a critique of the adaptationist programme. *Proc. Roy. Soc. Lond.* **B205**, 581–98.

Griffin, D. R. (1981). *The Question of Animal Awareness* (2nd edn). W. Kaufmann, Inc., Los Altos.

Hamilton, W. D. (1964). The genetical evolution of social behaviour, I & II. *J. theoret. Biol.* **7**, 1–52.

Hoffman, A. (1983). Paleobiology at the crossroads: a critique of some modern paleontological research programs. In *Dimensions of Darwinism* (ed. M. Grene), pp. 241–71. Cambridge University Press, Cambridge.

Holder, N. (1983). Developmental constraints and the evolution of vertebrate digit patterns. *J. theoret. Biol.* **104**, 451–71.

Hull, D. L. (1980). Individuality and selection. *Ann. Rev. Ecol. System.* **11**, 311–32.

Hume, D. (1779). *Dialogues Concerning Natural Religion*. Reprinted 1965 by Hafner Publ. Co., NY.

Huxley, J. S. (1932). *Problems of Relative Growth*. The Dial Press, New York.

—— (1954). The evolutionary process. In *Evolution as a Process*. (eds J. S. Huxley, A. C. Hardy and E. B. Ford). Allen and Unwin, London.

Huxley, L. (1902). *Life and Letters of Thomas Henry Huxley*, Vol. 2. Macmillan, London.

Krauskopf, K. (1953). *Fundamentals of Physical Science*. McGraw-Hill, New York.

Labov, J. B. (1981). Pregnancy blocking in rodents: adaptive advantages for females. *Am. Nat.* **118**, 361–71.

Leigh, E. G., Jr (1977). How does selection reconcile individual advantage with the good of the group? *Proc. Nat. Acad. Sci. (U.S.A.)* **74**, 4542–6.

May, M. T. (1968). *Galen on the Usefulness of the Parts of the Body*. Cornell University Press, Ithaca.

Maynard Smith, J. (1977). Parental investment: a prospective analysis. *Anim. Behav.* **25**, 1–9.

—— (1978). Optimization theory in evolution. *Ann. Rev. Ecol. System.* **9**, 31–56.

—— (1982). *Evolution and the Theory of Games*. Cambridge University Press, Cambridge.

—— (1983). Current controversies in evolutionary biology. In *Dimensions of Darwinism* (ed. M. Grene), pp. 273–86. Cambridge University Press, Cambridge.

Mayr, E. (1983). How to carry out the adaptationist program? *Am. Nat.* **121**, 324–34.

Medawar, P. (1974). A geometric model of reduction and emergence. In *Studies in the Philosophy of Biology. Reduction and Related Problems* (eds F. J. Ayala and Th. Dobzhansky), pp. 57–63. University of California Press, Berkeley.

Michod, R. E. (1983). Book review. *Am. Scient.* **71**, 525.

Midgley, M. (1979). Gene-juggling. *Philosophy,* **54**, 439–58.

Patten, B. C. & Odum, E. P. (1981). The cybernetic nature of ecosystems. *Am. Nat.* **118**, 886–95.

Popper, K. R. (1974). Scientific reduction and the essential incompleteness of all science. In *Studies in the Philosophy of Biology. Reduction and Related Problems* (eds F. J. Ayala and Th. Dobzhansky). University of California Press, Berkeley.

Queller, D. C. (1983). Kin selection and conflict in seed maturation. *J. theoret. Biol.* **100**, 153–72.

Raup, D. M. and Gould, S. J. (1974). Stochastic simulation and evolution of morphology – towards a nomothetic paleontology. *Syst. Zool.* **23**, 305–22.

Reichman, O. J. and Aitchison, S. (1981). Mammal trails on mountain slopes: Optimal paths in relation to slope angle and body weight. *Am. Nat.* **117**, 416–20.

Rhoades, D. F. and Bergdahl, J. C. (1981). Adaptive significance of toxic nectar. *Am. Nat.* **117**, 798–803.

Ridley, M. (1978). Paternal care. *Anim. Behav.* **26**, 904–32.

—— and Rechten, C. (1981). Female sticklebacks prefer to spawn with males whose nests contain eggs. *Behaviour,* **76**, 152–61.

Rothstein, S. I. (1982). Successes and failures in avian egg and nestling recognition with comments on the utility of optimality reasoning. *Am. Zool.* **22**, 547–60.

Shapiro, A. M. (1981). The pierid red-egg syndrome. *Am. Nat.* **117**, 276–94.

Simberloff, D. (1981). Community effects of introduced species. In *Biotic Crises in Ecological and Evolutionary Time* (ed. M. H. Nitecki). Academic Press, New York.

Sober, E. and Lewontin, R. C. (1982). Artifact, cause and genic selection. *Philos. Sci.* **49**, 157–80.

Stanley, S. M. (1975). A theory of evolution above the species level. *Proc. Nat. Acad. Sci. USA* **72**, 646–50.

Teilhard de Chardin, P. (1959). *The Phenomenon of Man.* Harper, New York.

Thornhill, R. (1984). Scientific methodology in entomology. *Florida Entomologist,* **67**, 74–96.

Tinbergen, M. (1963). On aims and methods of ethology. *Zeit. Tierpsychol.* **20**, 410–33.

Tribus, M. (1966). Micro- and macro-thermodynamics. *Am. Scient.* **54**, 201–10.

Van Valen, L. (1975). Group selection, sex, and fossils. *Evolution,* **29**, 87–98.

Vitt, L. J. (1981). Lizard reproduction: habitat specificity and constraints on relative clutch mass. *Am. Nat.* **117**, 506–14.

Vrba, E. S. (1983). Marcoevolutionary trends: new perspectives on the roles of adaptation and incidental effect. *Science,* **221**, 387–9.

—— (1984). What is species selection? *Syst. Zool.* **33**, 318–28.

Waddington, K. D. (1983). Pollen flow and optimal outcrossing distance. *Am. Nat.* **122**, 147–51.

Wade, M. J. (1979). The evolution of social interactions by family selection. *Am. Nat.* **113**, 399–417.

Wallace, B. (1985). Reflections on the still-'hopeful monsters'. *Q. Rev. Biol.* **60**, 31–42.

Wassermann, G. D. (1981). On the nature of the theory of evolution. *Philos. Sci.* **48**, 416–37.

Watson, P. J., LeBlanc, S. A., and Redman, C. L. (1971). *Explanation in Archaeology.* Columbia University Press, New York.

Watson, R. A. (1982). Book Review. *Q. Rev. Biol.* **57**, 221–2.

Williams, G. C. (1966). *Adaptation and Natural Selection.* Princeton University Press, Princeton.

—— (1975). *Sex and Evolution.* Princeton University Press, Princeton.

—— and Koehn, R. F. (1978). Genetic differentiation without isolation in the American eel, *Anguilla rostrata.* II. Temporal stability of genetic patterns. *Evolution,* **32**, 624–37.

Willson, M. F. and Burley, N. (1983). *Mate Choice in Plants.* Princeton University Press, Princeton.

Wilson, D. S. (1980). *The Natural Selection of Populations and Communities.* Benjamin/Cummings Publ. Co., Menlo Park.

Wimsatt, W. C. (1980). Reductionist research strategies and their biases in the units of selection controversy. In *Scientific Discovery: Case Studies* (ed. T. Nickles), pp. 213–59. D. Reidel Publ. Co., Boston.

Yodzis, P. (1981). The structure of assembled communities. *J. theoret. Biol.* **92**, 103–17.

—— (1984). The structure of assembled communities II. *J. theoret. Biol.* **107**, 115–26.

A geometric view of relatedness

ALAN GRAFEN

1. Introduction

The first objective of this paper is to introduce a geometric view of relatedness, as it is relevant to the evolution of social behaviour in the way pioneered by Hamilton (1963, 1964). The geometric view will suggest a new definition of relatedness. The exposition of this view will cover a number of facts about relatedness that have hitherto been available only to the mathematically competent (or at least adventurous) reader. The second objective follows on naturally from the first. It is to consider the status of Hamilton's rule as an evolutionary principle. There has been a steady trickle of interesting population genetical theory about the validity of Hamilton's rule and this will be briefly reviewed later. My main point will be that so far as the likely evolution of a character is concerned, this increasing body of theory confirms Hamilton's rule in a broad range of circumstances. The cases where Hamilton's rule does not apply will be explained with reference to the geometric view of relatedness.

In pursuit of these two aims, a number of other matters will come to our attention. An increasingly popular approach to population genetics, the covariance selection mathematics of Price (1970, 1972), will be explained and used in a new derivation of Hamilton's rule. A rough computation will be made of the frequency of unrelated individuals that happen to be as genetically similar to an animal as are its relatives of known degree. This is relevant to the suggestion of Hamilton (1971) and others that individuals may detect genetic similarity directly. We will also encounter, and eventually confront and solve, the 'paradox of inbreeding' of Seger (1981). The resolution will lead us to disagree with Jacquard (1974, p. 171), who suggested that relatedness is a measure of our information and not of anything real. If this relativism were true, then Hamilton's rule would be meaningless.

After that brief summary of my intentions, I wish now to give a fuller introduction to the paper. The readers for whom this introduction is intended have met the concept of relatedness and Hamilton's rule, and find them so unproblematic that they are surprised that any clarification, defence, or exposition is necessary. A fair sized literature, to which reference will be made later, deals with relatedness and Hamilton's rule, and its very existence is a good indication that there are problems with these ideas. However, this literature is mainly mathematical and I aim now to persuade the confident reader, using words only, that clarification, defence, and exposition are, after all, necessary for Hamilton's rule and the concept of relatedness. I will then give the plan of the rest of the paper, section by section.

Firstly, let us take the question whether relatedness is such an easy idea. A first attempt at showing that full, diploid sibs are related by one-half is to say that they share half their genes and that they do so because each gene in their parents has a 50 per cent chance of ending up in each sibling independently. However, we all share most of our genes, for, at least with the technology available to us a few years ago, the homozygosity in human populations was about 80 per cent (Cavalli-Sforza and Bodmer 1971). This means that two alleles chosen at random from the same locus from random human beings have an 80 per cent chance of being the same. If we really meant what we said about relatedness, we would conclude that 'unrelated' individuals are, in fact, closely related, with a relatedness of 0.8. The next step in the argument (Dawkins 1979) is to say that when we said 'share half their genes', we meant not only that the genes had to be identical, but also that they had to be identical by descent (Malécot 1969). If their parents are unrelated, then all identity by descent between the siblings must be by descent from their parents and our conclusion that siblings are related by one-half is vindicated.

Before proceeding, I should assure the reader that I do not doubt that the sibs are related by a half, but I am anxious that the argument that gets us there should be logically impeccable. When we ask if this idea of identity by descent can be taken seriously for our present purposes, we run straight into what Seger (1981) discussed under the heading 'The paradox of inbreeding'. The paradox is seen by first supposing that in the evolution of a particular locus, every mutation gives rise to a new allele. It then follows that any two identical alleles in the population now are descendants of the same, original, newly mutated allele. Therefore, any two identical alleles are identical by descent. (This conclusion applies in a weaker, but still vicious form if mutant alleles are not all unique.) We have come full circle and are faced again with knowing that sibs are related by one-half, but not being able to say why.

This problem with the meaning of relatedness is a crucially important point in trying to prove that Hamilton's rule is correct. (Hamilton's rule states that selection will favour an action by one animal that causes a loss to itself of c offspring, and a gain of b offspring to another animal to which it is related by r, provided $rb - c > 0$.) Two approaches have been taken. One is to show that Hamilton's rule works for specific kinds of relatives. The other is to prove that Hamilton's rule works for any kind of relative, using some mathematical definition of relatedness. The first method leads to greater mathematical respectability, while the second is biologically more general.

The mathematical definitions used in the second method do not refer to kin connections at all, but only to the fraction of genes shared by interactants. It follows that these definitions do not in themselves run up against the paradox of inbreeding, but the paradox will not go away. If we wish to apply Hamilton's rule using relatednesses derived from our knowledge of common ancestry, then we need to be able to make the connection between common ancestry and the relatedness defined by the fraction of genes interactants share, and there the paradox creeps in.

I hope that I have explained why relatedness is a problematic concept. The confident reader may also be surprised that mathematical models are built to try to prove or disprove Hamilton's rule. Its truth may seem sufficiently obvious, but again this is not so. In the simplest, central case, there is little doubt that the rule does apply. Equally, there are now many cases known where there is no doubt that Hamilton's rule does not apply. Charlesworth (1978) gave one good example in which Hamilton's rule fails, which I shall recount briefly. Nestlings that are full sibs have the opportunity to commit suicide to increase the survival of their fellow nestlings. What is the condition for a dominant allele to spread that causes nestlings to take this opportunity? How large must the increase in survival of their fellow nestlings be to justify this sacrifice?

The answer is that this allele can never spread. All the bearers of the altruistic allele commit suicide. It follows that there will be no copies of the altruistic allele next generation. The point of this example is that a naive application of Hamilton's rule would give the wrong answer. We need principles that tell us when Hamilton's rule will fail, as it does here. That is the point of trying to prove Hamilton's rule. The interest is focussed on what assumptions we need to make to prove it. A general proof with few assumptions makes Hamilton's rule a general rule; if only proofs with many special assumptions can be found, this means Hamilton's rule is of limited applicability. A biologist who applies Hamilton's rule in the field should be most worried about its scope of validity, for only if it is true for all sorts of genetic systems can he hope that it applies to the usually unknown genetic system that governs the character he is studying.

Relatedness, then, is a concept that needs explaining and Hamilton's rule needs proving. The second method of modelling Hamilton's rule mentioned above consists of finding the right concept of relatedness to make Hamilton's rule work and that is what I shall do here.

Section 2 contains the new derivation of Hamilton's rule and, on the way, an exposition of the covariance selection mathematics of Price (1970, 1972). It concludes with a new definition of relatedness, which is the main point of the section. Section 3 is, by contrast, not at all mathematical and explains the geometric interpretation of the new definition of relatedness. It also explains how this definition of relatedness is connected to ordinary notions of kinship and why it is the right definition to make Hamilton's rule work. This is the section that contains the first main message of the paper and readers who are positively afraid of equations should start with it. Other sections can be viewed as showing that different parts of this message are true. Section 2 proves that the new definition makes Hamilton's rule work. Section 4 proves that the geometric definition of section 3 is the same as the algebraic one of section 2. Section 5 proves that the geometric definition of section 3 really is connected to ordinary notions of kinship under certain assumptions, which it goes on to spell out and explain.

The last three sections are almost free of mathematics. Section 6 contains the second main message of the paper, that Hamilton's rule is a good evolutionary principle. It does this by arguing that the assumptions of section 5, while fairly restrictive for the application of Hamilton's rule to population genetics questions such as the rate of spread of particular alleles, are much less restrictive for the application of Hamilton's rule to broader evolutionary questions, such as the likely outcome of selection on a particular character. In the seventh section, I survey previous papers on Hamilton's rule and relatedness, and section 8 contains brief conclusions.

I have tried to make the paper intelligible to those to whom equations appear as a blur and to the extent that I have failed I apologize. As is often the case, the important conclusions and arguments are not mathematical, but a few essential points are. I hope that, provided these few essential points are taken on trust, the non-mathematically inclined reader can follow the important arguments and understand the important conclusions.

2.   A new derivation of Hamilton's rule

This section contains a new derivation of Hamilton's rule, using the covariance selection mathematics of Price (1970, 1972). Price's method is also explained in some detail. Some previous derivations have used Price's method (Hamilton 1970, 1975; Seger 1981; Uyenoyama et al. 1981) to varying degrees. Applications of Price's method will be reviewed later in this section. Here I give a derivation that allows each individual to have its own ploidy, and whose formulation covers both single allele and multi-allele, and single-locus and multi-locus models. These extensions are natural within Price's method and their value is that they unify in one treatment results that would otherwise have to be proved separately. For the purposes of this paper, they mean that our definition of relatedness will be appropriate for arbitrary ploidies and any number of alleles. Other derivations will be discussed in section 7, when ideas to be developed later will allow more fruitful comments to be made.

What exactly, then, is the rule we are setting out to derive? It is a rule that tells us whether natural selection will favour any social action, where by a social action we mean an action that has consequences for the number of offspring of the animal performing it and also for the number of offspring of some other animal of the same species. According to the rule, we need to know only three numbers to decide if selection will favour a social action. Firstly, the effect of the action on the actor's number of offspring; because of the special interest in altruistic actions, this is conventionally measured as the decrease in the actor's number of offspring, is called a cost and denoted c. Secondly, the effect of the action on the recipient's number of offspring, conventionally measured as an increase, is called a benefit and denoted b. Thirdly, a quantity known as the relatedness of the actor to the recipient, denoted r. Hamilton's rule states that the social action is favoured by natural selection if

$$rb - c > 0. \tag{1}$$

The rule was first derived by Hamilton (1963, 1964) and it has a very simple biological interpretation. The actor values one offspring of the recipient as a certain fraction of one offspring of its own. That fraction is the relatedness. The remarkable property of eqn (1) is its simplicity as a summary of, or prediction about, the results of quite complicated population genetics models. The dominance of the alleles involved is not mentioned in eqn (1), nor are the number of loci and the ploidies of actor and recipient. In many ways it looks too good to be true. Our aim in deriving Hamilton's rule, when we are not quite sure how relatedness is to be defined, is to find a condition that tells us whether a social action is favoured by natural selection, then to see if we can choose a definition of r to make this condition equivalent to eqn (1). This definition is the one we will adopt. Of course, we must then examine its properties to see what resemblance it bears to our notions of kinship and common ancestry.

This is a good place to introduce a distinction, made by Crozier (1970), between relatedness and relationship. We may explain r in eqn (1) in two ways. The first is to say that it measures the genetic similarity between donor and recipient, the extent to which they possess the same genes (Crozier's 'relatedness'). The second is to say that r is a measure of common ancestry and can be computed from a family tree (Crozier's 'relationship'). Now genetic similarity has many possible causes, of which common ancestry is only one. Genetic similarity makes Hamilton's rule work, but it is common ancestry that we are likely to know when we wish to apply it. For this and other reasons we shall come across in due course, Hamilton's rule is most useful when the genetic similarity is caused solely by common ancestry. In section 6, we shall see that common ancestry causes genetic similarity of a very special kind. The definition of r we uncover later in this section will be a measure of genetic similarity. Its connection with common ancestry is the topic of section 5. While the substance of the distinction is very important, it is confusing to use such similar words for concepts to be contrasted. In this paper, I have used the word relatedness for both concepts where confusion is unlikely, and explicitly refer to genetic similarity or common ancestry where necessary.

In the next subsection, Price's method is developed and explained and its uses reviewed. The following subsection uses Price's equation to derive Hamilton's rule.

PRICE'S COVARIANCE SELECTION MATHEMATICS

Now we turn to Price's method, which is a pleasing way of doing population genetics (Price 1970, 1972). I give here an account of one fairly general use of Price's method which I have found useful. Other accounts of aspects of Price's method are given by Hamilton (1975), Seger (1981), and Wade (1985). Assume discrete generations. We begin by taking all the individuals in one generation and indexing them, that is to say, giving each of them a number rather as houses in a street are given numbers for ease of reference. We then want to measure four things about each individual, considered as a potential parent. They will be four numbers and every potential parent will have its own set of these four numbers. For the ith individual, they will be denoted:

l_i the ith individual's ploidy;

w_i the number of successful gametes of the ith individual, per haploid set;

p_i the ith individual's 'p-score' (see below for definition);

$p_i + \triangle p_i$ the average p-score of the ith individual's successful gametes.

Let's take these in turn. An individual's ploidy is the number of haploid sets of chromosomes there are in its genome: this is one for a haploid, such as most male hymenopterans and human gametes; two for a diploid, such as hymenopteran females and humans; three for a triploid, such as some plant endosperm, and so on. We will be very general in our approach and allow each individual to have any ploidy. The usual circumstance is that all males have one ploidy and all females have another. For problems concerning other genetic entities, such as gametes or plant endosperm, it is useful to have a general result.

The number of successful gametes per haploid set is just what it says. A successful gamete is one that contributes to an offspring in the next generation at whatever stage we decide to count them. The reason for dividing by the ploidy of the individual parent is its significance in the theory. We want to know the average success of a haploid set. The sole parent of a haploid with two offspring is better represented genetically in the second generation than is one of the two parents of a diploid with three offspring.

The third and fourth numbers, p_i and $p_i + \triangle p_i$, lie at the heart of Price's method, and to explain them I must explain what I mean by a p-score. The p-score can be anything that an offspring inherits by averaging together the gametic contributions of the parents. The simplest example is where the p-score of an entity (individual or gamete, or group of individuals or gametes), is defined as the frequency within it of one particular allele. Thus, in a haploid gamete, the p-score is either 0 or 1. In a diploid individual, the p-score is either 0, 0.5, or 1. In a population of 50 diploids, the p-score can be any number out of 100. Let us check that this kind of p-score satisfies our requirements. If the p-score of the gametes that made you were both 0, then neither had the allele, so neither do you, so your p-score is 0. If one gametic p-score was 0, and the other was 1, then you are a heterozygote and your p-score is 0.5. If both p-scores were 1, then you have two copies of the allele and your p-score is also 1. In every case, your p-score is the average of the p-scores of the gametes that made you.

A more interesting choice for a p-score involves all the alleles at one locus. We can give every allele at this locus a separate number, any number we choose. Then the average of the numbers assigned to the alleles an entity possesses is a p-score. For example, suppose we have three alleles, A, B, and C, and we assign them the values 1, 3, and 11, respectively. A B gamete has a score of 3, an AC diploid has a score of 6, and an $AABC$

tetraploid would have a score of 4. The score of an entity is the average of the scores of the gametes that made it and so this score qualifies as a p-score.

One final extension will complete our range of p-scores. The genes in the previous example were alleles at the same locus. This was an unnecessary assumption, because even if they were genes at different loci, the defining property of a p-score would have been satisfied. Thus, the most general kind of p-score for a haploid entity is the sum of arbitrary values attached to any number of alleles at any number of loci. For other entities, it is the average p-score of the haploid sets that make it up.

Our derivation will proceed by following the evolutionary change in a p-score; because we will not specify which p-score, our conclusions will be true for any p-score. The range of p-scores for which Price's method works means that the same equations will serve for a two allele, single-locus model, when the p-score is interpreted as the frequency of one allele; for a polygenic model, in which the p-score is interpreted as the summation of the small allelic effects that combine to produce the polygenic character and, indeed, for a whole range of intermediate models. The additive genetic value of any character (Falconer 1981) can be represented as a p-score. All this is achieved at the same time as allowing each individual to have its own ploidy.

Having defined what a p-score is, the third and fourth numbers in the list above can now be explained. p_i is the p-score of the ith individual and the average p-score of its successful gametes is $p_i + \triangle p_i$. We write the p-score of the ith individual's successful gametes as $p_i + \triangle p_i$, because there is a presumption that an individual's gametes have the same gene frequencies as the individual has, and so $\triangle p_i$ will, on average, be 0. The randomness of meiosis will often make it non-zero in a particular case and meiotic drive (for an explanation see Wright 1968, 1969, 1977, 1978) would make it non-zero on average. Let us agree to ignore the effects of meiotic drive, and so assume where necessary that on average $\triangle p_i$ is 0.

There are two more notational devices to mention before we get down to work. The first is that a symbol without a subscript represents an average over all the haploid sets that make up the population. Thus, p is the average p-score of all the haploid sets in the parental population. Alternatively, we can think of it as the average p-score taken over individuals, with each individual weighted by its ploidy. The second device is that a prime (') denotes the value among the offspring, so that p' represents the average p-score of all the haploid sets that make up the offspring. p' is also the p-score of the successful gametes of the parents and it is this equivalence that we now exploit.

We now go on to derive Price's basic equation. The number of successful gametes of the ith individual, and the sum of all the p-scores of those successful gametes, are

$$l_i w_i$$

and

$l_i w_i (p_i + \triangle p_i)$.

So, using the $\sum$ notation to denote summing over i, that is adding up these values for all the individuals in the population, we obtain

$\sum l_i w_i$

and

$\sum l_i w_i (p_i + \triangle p_i)$.

However, if these are the number of successful gametes and the summed scores, respectively, for all successful gametes, then the average p-score of the successful gametes must be given by

$$p' = \frac{\sum l_i w_i (p_i + \triangle p_i).}{\sum l_i w_i} \qquad (2)$$

To produce Price's equation from this we need to use two statistical notions, the expectation and the covariance. The expectation is just an average, but I use the word to explain why I will use the symbol **E** to denote an average. It is an average of the sort described above, over haploid sets. The covariance of two quantities is the average product minus the product of the averages and will be denoted **Cov**. Putting these definitions into symbols gives us

$$p = \mathbf{E}p_i = \frac{\sum l_i p_i,}{\sum l_i}$$

$$w = \mathbf{E}w_i = \frac{\sum l_i w_i,}{\sum l_i}$$

$$\mathbf{E}(w_i \triangle p_i) = \frac{\sum l_i w_i \triangle p_i,}{\sum l_i}$$

$$\mathbf{Cov}(w_i, p_i) = \mathbf{E}(w_i p_i) - \mathbf{E}(w_i)\mathbf{E}(p_i)$$

$$= \frac{\sum l_i w_i p_i}{\sum l_i} - \frac{\sum l_i w_i}{\sum l_i} \frac{\sum l_i p_i.}{\sum l_i}$$

With these definitions we can rewrite eqn (2) as follows, letting $\triangle p = p' - p$, in keeping with our notational conventions,

$$w \triangle p = \mathbf{Cov}(w_i, p_i) + \mathbf{E}(w_i \triangle p_i). \qquad (3)$$

This is Price's equation, and it expresses in a very general and precise way an obvious truth. The left-hand side, $w\triangle p$, is the mean fitness of the parental population multiplied by the change in mean p-score. As the mean fitness is always positive, eqn (3) says that we can deduce the direction of evolutionary change by finding the sign of the right-hand side. The covariance term is positive if an individual's fitness is positively correlated with its p-score and negative if that correlation is negative. This just tells us that if a high p-score is correlated with a high number of successful gametes, then it will tend to be increased by selection. We could plot number of successful gametes against p-score, representing each individual in the population as one point. If the best fitting straight line (using ploidies as weights) slopes up, then the covariance term is positive; if the line slopes down, then the covariance term is negative.

The second term on the right-hand side of eqn (3) is the gametic discrepancy term, representing the effect of the difference between an individual's own p-score and the p-score of its successful gametes. If we were studying random drift, then this term would be important. We have already agreed to neglect meiotic drive and so the gametic discrepancy term is on average 0. We now further assume the population is large enough for us to be able to ignore random fluctuations in this term. We will deal only with the covariance term.

The essentials of Price's method have now been derived and the reader anxious to use it should skip ahead. The rest of this subsection is an aside about Price's method. Firstly, following Price (1972), Hamilton (1975), and Wade (1985), I show how powerful a tool Price's method can be in analysing higher levels of population structure, but expand a point made by Hamilton (1975) about a complication in interpretation that is sometimes missed. Secondly, I review the uses to which Price's method has been put.

The extension to population structure is done simply by changing the interpretation of the symbols we have used. Suppose the population is divided into groups and that i indexes groups, not individuals. Then l_i is the number of haploid sets in the ith group, p_i is the average p-score of the haploid sets in the ith group, and $p_i + \triangle p_i$ is the average p-score of the successful gametes of the ith group. Equation (3) is still true, by the same argument, for this new interpretation of the symbols. The covariance term now represents the effect of selection on groups and the second term is no longer gametic discrepancy. Rather, it is the effect of the difference between the p-score of the groups and the p-score of the successful gametes of the group. Part of this is the effect of individual selection within groups, by which individuals with, for example, high p-scores have a higher fraction of their group's reproduction than individuals with low p-scores.

This decomposition can be expressed formally, using a second level of subscripting. Let us use g to index groups, and gi to index the ith individual within the gth group. The expectation and covariance may also be subscripted, because we can ask for the average or covariance between groups, and this is what $\mathbf{E}$ and $\mathbf{Cov}$ stand for; we can also ask for the average within the gth group, and this is what $\mathbf{E}_g$ and $\mathbf{Cov}_g$ represent. The decomposition is as follows:

$$w \triangle p \; = \; \mathbf{Cov}(w_g, p_g) + \mathbf{E}(w_g \triangle p_g)$$

$$= \; \mathbf{Cov}(w_g, p_g) + \mathbf{E}\{\mathbf{Cov}_g(w_{gi}, p_{gi}) + \mathbf{E}_g(w_{gi}, p_{gi})\}.$$

The second line is obtained by noticing that the second term in the first line is the average over groups of an expression of the form $w \triangle p$ and so substituting for it using eqn (3). This recursive scheme could be repeated by substituting for the third term in the second line, and so on, as discussed by Hamilton (1975).

It is tempting to suppose that if the within group covariance is negative, then individual selection is acting against the trait; and if the between group covariance is positive, then group selection is acting in favour of the trait; and that in such a case the trait is altruistic in that it favours the group, but harms the individual who expresses it. It is important to realize that this is false, if we interpret altruistic as originally defined by Hamilton (1964). The reason is that the between groups covariance contains, as Hamilton (1975) put it, 'a group selection component which is not 0, but which is bound in unchanging subordination to the individual selection component'. The between group covariance must be greater than this subordinate component for the trait to be altruistic in Hamilton's sense.

It is not hard to understand the source of this subordinate component. Suppose a trait affects only the bearer's fitness and is simply advantageous. Then the within group covariance is positive, because bearers within the group have more offspring than non-bearers. However, the between group covariance is also positive, because the groups with more bearers have more offspring than groups with fewer bearers. This positive between group covariance is not the result of any help given by the individual to other members of the group, but occurs because an individual possessing the allele is himself a fraction (one over the group size) of the group. This is Hamilton's 'subordinate component'. In cases where there is help given to fellow group members, its influence could be measured by subtracting the subordinate component from the between group covariance. Better in my opinion is to follow Hamilton (1975) in analysing the trait in terms of Hamilton's rule.

I now turn to a brief review of the influence of Price's covariance selection mathematics. The essence of Price's method, as he expressed it in his 1972 paper, is the naming of individuals rather than the naming of genotypes. This means that a subscript refers to an individual rather than to a genotype or to a gene. This led naturally, in Price's hands at least, to the application of the statistical notions of expectation and covariance in his population genetics models. The method he devised has not been much used, but two results he derived from it are sometimes cited.

The first result for which he is often cited is that the change in a character resulting from selection is equal to the genetic covariance of that character with fitness. This result was derived in a very different way by Robertson (1966) and named by him (1968) the 'secondary theorem of natural selection'. Crow and Nagylaki (1976) develop this theorem further. For Robertson, this result was important because it allowed the effect of selection on a character to be estimated from known quantities, no matter what incidental selection might be simultaneously carried out. This incidental selection could arise either because of a correlation of characters within the population, or just because there were also other selective forces at work. Other methods depended on the selective force studied being the only one at work and on the absence of correlation with other characters relevant for fitness. Price values the result for its conceptual clarity and generality of expression.

The second result for which Price is cited is the hierarchical decomposition of a population variance in a character, or population covariance between characters, in the way we have just seen. The variance (or covariance) has a within individual component, then a within group component, then a within deme component, and so on, to as many levels as desired. The decomposition is a general one, allowing individuals to have different ploidies, groups to have different numbers of individuals, and so on. Wright (references in Wright 1969) had already suggested this kind of decomposition for his F-statistics. Price used this second result as an illustration of the power of the first.

The only papers to my knowledge that use Price's method at any length, as opposed to citing one of these two results, are Hamilton (1970, 1975), Seger (1981), Wade (1985), and this paper. I know from as yet unpublished work of my own, and from comments of others who have used it, that the method is extremely useful over and above these two results, and I expect that there will soon be a number of papers that put the method itself to good use.

THE DERIVATION OF HAMILTON'S RULE

I now return to the main business of the section, deriving Hamilton's rule, and to our standard interpretation of i as indexing individuals. We have obtained Price's equation, which describes the evolutionary change in any character. The next step is to model the social interactions that are the subject of Hamilton's rule, and in particular, to say how the fitness of an individual depends on its phenotype and on the phenotypes of the individuals with which it interacts. In mathematical terms, we wish to model w_i. Suppose pairwise interactions take place, in such a way that one individual, the actor, has an opportunity to help another, the potential recipient. If the act is committed, we suppose that the actor's number of offspring is decreased by c, while the recipient's number of offspring is increased by b. (b and c are more precisely the changes in the number of haplotypes supplied in successful gametes to offspring. In the usual case, a parent provides one gamete containing one haplotype for each offspring.) Let m_i be the number of interactions in which the ith individual is the actor and n_i the number in which it is the potential recipient.

In order to know how many offspring an individual has, we need to know what happens in these interactions. On what fraction of the m_i occasions on which it was the actor did the ith individual commit the social act? Let it be

h_i. On what fraction of occasions on which the ith individual was the potential recipient did the actor commit the act, and thus the ith individual receive the benefit? Let it be y_i. I will refer to h_i as the phenotype of the ith individual, because h_i summarizes its actions. Similarly, y_i is the mean phenotype of the actor on the occasions on which the ith individual is the potential recipient. The net effect of all these social interactions on the number of offspring of the ith individual is

$$(n_i y_i)b - (m_i h_i)c.$$

Now this is the difference made by these social acts, but it is possible that the baseline fitness, to which this difference must be added, varies from individual to individual. The most obvious reason is that the baseline fitness of males and females may be different. If there are different ploidies within one sex, then the baseline fitness may be different for different ploidies. Finally, the baseline fitness may vary because selection of other characters is going on in the population or simply through random variation in number of offspring. Let the ith individual's baseline fitness be f_i. Then its number of offspring is

$$f_i + n_i y_i b - m_i h_i c.$$

w_i is the number of offspring of the ith individual, divided by its ploidy, and so we arrive at the following:

$$w_i = \frac{1}{l_i} (f_i + n_i y_i b - m_i b_i c). \tag{4}$$

This is the required model of w_i.

The next step is to combine our model of social interactions with Price's equation, so that we can look at the effect of social interactions on the systematic evolutionary changes in a p-score. We do this by using eqn (4) to substitute for w_i in eqn (3) without its last, gametic discrepancy, term. This gives

$$w \triangle p = \mathbf{Cov}(p_i, w_i)$$

$$= \mathbf{Cov}(p_i, \frac{1}{l_i} \{f_i + n_i y_i b - m_i h_i c\})$$

$$= \mathbf{Cov}(p_i, \frac{f_i}{l_i}) + b\mathbf{Cov}(p_i, \frac{n_i y_i}{l_i}) - c\mathbf{Cov}(p_i, \frac{m_i h_i}{l_i}). \tag{5}$$

The second step follows because covariances are distributive over addition.

The first of the three covariance terms is between the p-score and baseline fitness, and represents the effect on evolutionary change in the *p*-score of forces other than the social interactions. It is the remaining two terms that we will study further. They are to be rearranged. At the moment they are covariances where each data point is an individual. Hamilton's rule is phrased in terms of occasions when help is or is not given, and the rearrangement is to convert the covariances across individuals into covariances across occasions. Our focus shifts from a list of individuals to a list of occasions. Just as each individual had a list of numbers associated with it, so each occasion has a list of numbers. Let us index occasions by *j*, just as we indexed individuals by *i*. Thus, we will speak of the *j*th occasion. We define the total number of occasions as *J*. The numbers for each occasion are

D_j the *p*-score of the actor on the *j*th occasion;
R_j the *p*-score of the potential recipient on the *j*th occasion;
H_j the phenotype of the donor on the *j*th occasion, that is 1 if the act is committed and 0 if it is not. (h_i is the average of H_j for those occasions on which the donor was the *i*th individual.)

The expressions I will form from these variables are not strictly covariances, though they are covariance-like. I use the symbol **K** to represent them, and they are defined by:

$$\mathbf{K}(H_j, D_j) = \frac{1}{J} \ \Sigma H_j (D_j - p)$$

and

$$\mathbf{K}(H_j, R_j) = \frac{1}{J} \ \Sigma H_j (R_j - p).$$

Thus, the first **K** is the average value across occasions of the product of the donor's phenotype and the donor's deviation from the mean *p*-score of the population. The second is the average across occasions of the product of the donor's phenotype and the recipient's deviation from the mean *p*-score of the population.

The **K**'s would be covariances if *p* were the average *p*-score of actors in the first expression and the average *p*-score of potential recipients in the second. Instead, in both cases it is the average *p*-score of the population as a whole. This will be relevant later.

In order to assert the algebraic relationship between the covariances across individuals and the **K** forms across occasions, let α be the number of occasions in a generation divided by the number of haploid sets in the parental population. That is,

$$\alpha = \frac{\Sigma m_i}{\Sigma l_i} = \frac{\Sigma n_i}{\Sigma l_i} = \frac{J}{\Sigma l_i}.$$

The following identities are easily proved by expanding the covariances and **K** forms:

$$\mathbf{Cov}(\frac{m_i h_i}{l_i}, p_i) = \alpha\mathbf{K}(H_j, D_j)$$

$$\mathbf{Cov}(\frac{n_i y_i}{l_i}, p_i) = \alpha\mathbf{K}(H_j, R_j).$$

We can use these equations to substitute in eqn (5), and find that, for a p-score uncorrelated with baseline fitness,

$$w\triangle p = \alpha b\mathbf{K}(H_j, R_j) - \alpha c\mathbf{K}(H_j, D_j)$$

and so, providing $\mathbf{K}(H_j, D_j) \neq 0$,

$$w\triangle p = \alpha\mathbf{K}(H_j, D_j)\left\{\frac{\mathbf{K}(H_j,R_j)}{\mathbf{K}(H_j, D_j)}\ b - c\right\} \qquad (6)$$

This is the end of the derivation. Equation (6) is the expression we have been looking for; let us see why. The p-score increases from one generation to the next if $w\triangle p$ is positive and decreases if it is negative. The sign of $w\triangle p$ can be found from the signs of the three terms on the right hand side of eqn (6). α is positive by definition. The second term is positive for p-scores that are positively associated with committing the social action when the opportunity arises; let us restrict attention to those p-scores. Whether the p-score then increases or decreases depends on the sign of the third, bracketed term and it is to this we now turn.

If we allow ourselves to represent the ratio of **K** forms by one symbol and go so far as to use the symbol r, then we can rewrite the condition that the decisive third term is positive, so that the p-score increases, as

$$rb - c > 0.$$

However, this is exactly Hamilton's rule, as stated in eqn (1). We have proved Hamilton's rule, given that we are prepared to define r to be the ratio of **K** forms in the third term of eqn (6); that is

$$r = \frac{\Sigma H_j(R_j - p)}{\Sigma H_j(D_j - p)}. \qquad (7)$$

The problem now is that having defined r in this way, as a measure of genetic similarity (Crozier's relatedness), we cannot assume that it has any connection with kinship (Crozier's relationship). Equation (7) is a particular way of measuring genetic similarity and its differences from previously suggested measures will be discussed in section 7. The most useful property of relatedness, as usually understood, is that it expresses a connection between two individuals that can be computed from ancestry.

Equation (7), on the other hand, looks as if it could be different for different p-scores and for different characters, which would be a serious inconvenience in applying Hamilton's rule. However, all we can do is to explore the properties of r as defined by eqn (7) and hope for the best; for that is the definition which makes Hamilton's rule work and so we are stuck with it, however inconvenient it may turn out to be. Much of the rest of the paper is concerned with exploring the consequences of defining relatedness by eqn (7).

I end this section by reviewing it. We set out to find a definition of relatedness that would make Hamilton's rule work and we found one. In the process, we encountered the very powerful technique of Price's method. Our derivation of Hamilton's rule is valid for populations in which different members have different ploidies, for asexual, bisexual, and trisexual populations, for varying numbers of social encounters per individual, and for cases where the interacting individuals are not genetically representative of the population. It is true for populations with any kind of geographical structure and any kind of inbreeding. This generality will prove to be somewhat illusory, however, owing to complications in the interpretation of r. Hamilton's rule is true for any p-score on which selection acts only through the social behaviour modelled, but for other p-scores it tells us the effect of the social behaviour on the direction of selection of the p-score. (Naturally, it cannot predict changes in p-scores that are the consequence of processes we have not modelled!) In the next section we start from quite different considerations to arrive at a way of measuring genetic similarity that turns out to be intimately connected with the definition of relatedness discovered in this section. Succeeding sections explore implications of this coincidence.

3. A geometric view of relatedness

Leave aside the newly discovered formula for r, to which we shall return, and concentrate instead on genetic similarity, which is what we intend relatedness to be a measure of. There are many senses of similarity, and to explain the particular kind of genetic similarity I have in mind in this section, it helps to have a picture. Figure 1 is a very simple picture indeed. The line represents the frequency of one particular gene, where one end is 0 and the other end is 1. We can represent individual animals on this line according to the frequency of the gene they possess. A haploid creature must lie at one end of the line or the other, because it either has the gene or it doesn't. A diploid creature must sit at one of the ends if it is a

0   1

Fig. 1. A line representing a gene frequency, on which various entities can be represented. A haploid either has the gene or doesn't, and so its gene frequency is 0 or 1, and must lie at one end or the other of the line. A diploid's gene frequency is either 0 or 1 for a homozygote, or 0.5 for a heterozygote.

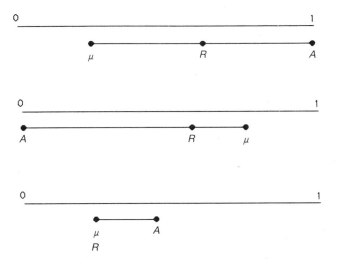

Fig. 2. Three examples of possible relationships between the gene frequencies of the population (μ), the actor (A) and the potential recipient (R). The relatednesses are, in order starting from the top, 0.5, 0.25, and 0. The population mean is not the same in each example.

homozygote, or in the middle if it is a heterozygote. We can also use a point to represent the average gene frequency in a population and, if the population is large, we can think of the population mean as lying at any position on the line.

Consider three points on the line. The point designated μ is the population mean, the point A is the actor, and R is the position of the average gene frequency of the potential recipients of the actor. This simple diagram contains all the necessary ingredients for the geometric view of relatedness, which is as follows. Take the line that runs from μ to A. Then the relatedness of the actor to the potential recipients is the fraction of the way along the line that R is found. If R is at the beginning of the line, at μ, then the relatedness is 0. If R is half-way in between the two points, then the relatedness is 0.5. If R is at the same point as A, then the relatedness is one. Some examples are illustrated in Fig. 2.

This is the geometric view of relatedness. I will develop it in a number of ways. I shall first explain how it can be extended into more dimensions so that it becomes a general picture of how whole genomes are related, rather than just of the presence of one allele at a single locus. I will then say how it relates to the more obvious ideas of relatedness such as sibship and other ancestral links, and thirdly why it is the right concept of relatedness to use in Hamilton's rule. These points can be seen as the formalization of the intuition that Hamilton's rule is obvious (at least in retrospect). In the next section I will make the connection between this simple picture of relatedness and the definition of r adopted in the previous section.

First the extra dimensions, to represent more than one allele, and more than one locus. The essentials of the picture, the points μ, R, and A stay

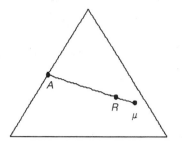

Fig. 3. The μ–R–A line in a triangle representing the allelic frequencies at a locus with three alleles. The relatedness is the fraction of the way from μ to A at which R is found.

the same: only the background changes. The simplest example is a locus with three alleles. The genotype at this locus can be represented as a point inside a triangle such as Fig. 3. The distance from each line represents the fraction of the corresponding allele. Haploid creatures must sit at one of the three corners. Diploid homozygotes also sit in a corner, but diploid heterozygotes sit at the midpoint of one of the sides. A triploid organism with one of each allele can sit in the middle of the triangle. Again, the mean of the population can be represented in the diagram and, if the population is large enough, it can be practically anywhere in the triangle.

Now think of the three points μ, R, and A, and in particular of the line from μ to A. Later a whole section is devoted to the possibility that R does not lie on this line, but assume for the moment that R does lie on the μ–A line. Again the relatedness is the fraction of the way along the line from μ to A at which R is found. The example can be extended to any number of alleles at one locus by imagining a tetrahedron for four alleles, then a tesseract for five, and so on.

Figure 4 shows a different extension to two dimensions. This is a square,

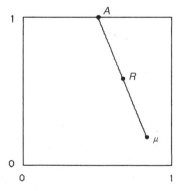

Fig. 4. The μ–R–A line in a square representing the frequency of one allele at each of two loci.

in which each axis corresponds to a different locus, each locus having two alleles. One point in the square represents the frequency of the alleles at each of two loci. Again individuals can be represented as points. Haploids sit at a corner. Diploids sit at a corner if they are double homozygotes, on the middle of an edge if they are homozygous at one locus and heterozygous at the other, and right in the middle of the square if they are double heterozygotes. The population can be represented by a point in the square, and so once again we can have the three points, μ, R, and A. Once more, the relatedness is the fraction of the way along the line from μ to A at which R is found. This same idea can easily be extended to any number of alleles at any number of loci. We can imagine a space in which one point represents the whole genome of an organism and the same three points could be found, and the same computation of relatedness made. μ, R, and A are what Jacquard (1974) called 'genic structures'.

Now what does this geometric picture of relatedness, of genetic similarity, have to do with ordinary notions of common ancestry, such as being someone's sister or second cousin? Suppose you are diploid, and you know your own position in genetic space and the position of the population mean as well. How can you compute the position of your offspring? Any particular offspring is at a particular place, of course, but where on average do you expect an offspring to be? Let's assume panmixia and an infinitely large population. One allele at each locus in your offspring comes from you and one from your mate. So your offspring are on average half-way between you and your mate in genetic space. Your mate is on average at the population mean, since this is the meaning of the assumption of panmixia. It follows that your offspring are, on average, half-way between you and the population mean. What then, in the geometric view, is your relatedness from you to your offspring? The answer is a half. Offspring themselves may have offspring, and their position will, by the same argument, be half-way between your offspring's position and the population mean. They are therefore three-quarters of the way from you to the population mean, and so they are one-quarter of the way from the population mean to you. You are therefore related to them by one-quarter.

A similar argument can be used for a sibling. Each of my two alleles at a locus came from a different parent, and so each allele had a separate 50 per cent chance of also being transmitted to my sibling, who shares both parents. So there's a 25 per cent chance of each of four possible outcomes: we share both alleles, only the paternal allele, only the maternal one, or neither. So the average fraction of alleles shared by copying from parents is a half and the average position of the shared part is my own position in genetic space. The unshared fraction of my sibling's genotype is filled with genes about which, because of the assumption of panmixia, I can assume that they are drawn at random from the population. Hence, the unshared fraction of a half has its average position in genetic space at the population mean. My sibling is on average half me, half the population mean, and so my sibling's average position is half-way between me and the population mean. So my relatedness to him is one-half.

Similar arguments can be constructed for other relationships. The point is not that these methods of computing relatedness are novel, far from it. Charnov (1977) applied them algebraically in his derivation of Hamilton's rule. Rather, the point is how easily the methods fit into the geometric view of relatedness. The picture is a helpful way of seeing how the arguments work.

Turning from how the geometric view accords with the concepts of common ancestry, we move on to the question of what connection it might have with the evolution of social behaviour. The mechanism of evolution is changes in gene frequency. If an individual at μ, the population mean, reproduces, then the gene frequencies are not changed because the addition to the offspring population has the same gene frequency as the adult population. On the other hand, if another individual, say the actor at A, reproduces, then this moves the offspring population slightly from μ to A, perhaps only a little way, but if everyone near A reproduces more than average, then this will cause the changes in gene frequency that underlie evolution.

Take A's view. Reproduction by an individual at μ, or equal reproduction by members of a group whose mean position is at μ, is irrelevant. However, someone else at A reproducing has exactly the same effect on gene frequencies as if A itself reproduced. If an individual half-way in between produces two offspring, then this will have the same effect on gene frequencies as if someone at μ had one offspring and someone at A had the other. Similarly, if an individual whose position in genetic space was one-eighth of the way from μ to A had eight offspring, then this would have the same effect on gene frequencies as if someone at μ had seven offspring and someone at A had one. The general drift must now be clear.

A physical analogy can be made. The process is rather like placing weights on a rod with a fulcrum at μ. To produce the same turning moment as one offspring at A, we would need two offspring at half the distance from μ to A, or four offspring at a quarter the distance from μ to A, and so on. The same turning moment corresponds to the same strength of effect on the gene frequency of the next generation. Hamilton (1971) used the analogy of different concentrations of liquid for the same purpose.

So if an individual reproduces who is a fraction λ of the way from μ to A, we can in our imaginations divide the set of offspring into two parts. The first, a fraction $1-\lambda$ of the offspring, at μ, and the second part, a fraction λ of the offspring at A. The combined average position of these two parts of the set of offspring is the same as the position of the original offspring, and so the effect on the gene frequency changes will be the same. However, the existence of the fraction at μ will not contribute to changes in the population gene frequency, because it is equal to the old population mean. The fraction at A, on the other hand, will change it and will change it to the same extent as if A had reproduced, but had only a fraction λ times as many offspring. Hamilton (1964) called the fraction $1-\lambda$ the 'diluting factor', because it does not change the direction of evolution, but does slow it down.

These arguments all show the same thing. That A should value R's reproduction as a certain fraction of its own, because they would have the same effect on gene frequencies as a certain fraction of its own. Further, that fraction is the distance along the line from μ to A at which R can be found.

My hope is that the reader is now convinced that the geometric view of relatedness is simple, is connected in a straightforward way with ordinary notions of common ancestry and is plausibly the right kind of concept to make Hamilton's rule work. Before going on to the next section, though, there is one complication that should be mentioned because we will not return to it. It is the possibility of negative relatedness, whose relevance to the evolution of social behaviour was first explained by Hamilton (1970, 1971).

Negative relatedness means that one individual is genetically less similar to another than it is to a random member of the population. It has an obvious geometric interpretation. R is on the line that passes through μ and A, but it is not in between those two points. Rather it is on the other side of μ from A, as in Fig. 5. R deviates from μ in the opposite way to that in which A deviates from μ. The consequence of this is simple enough. A values R's offspring negatively and will be prepared to give up offspring of its own to prevent R reproducing. This is what Hamilton called spite (Hamilton 1964, 1970, 1971; Knowlton and Parker 1979).

We can also say a little about how negative relatedness might arise and the simplest plausible way involves a small population size. μ is the population mean and A is one of the individual in the population. If we ask what is the relatedness of A to a random member of the population, the answer is immediate: it is zero because μ is at a fraction zero of the way from μ to A. However, what is the relatedness of A to an individual chosen randomly from the other members of the population, excluding the possibility that it might be A itself? The answer is illustrated in Fig. 5. Let N be the population size. Take R to be the average position of the members of the population excluding A. μ is the mean of R weighted by $N-1$, and A weighted by 1. It follows, as Hamilton (1971) first showed, that the relatedness of A to R is $-1/(N-1)$. This is negligible if the population is not very small, but is the basis for the models of spite cited above. The alert reader may have noticed that a small population affects our calculations of relatedness to relatives as well, for similar reasons, but

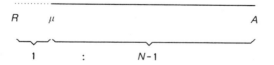

Fig. 5. The relatedness of an individual to the rest of the population is negative, because R lies on the other side of μ from A. The relative lengths of the line segments are 1:N-1 because μ is the average of R weighted by the N-1 other members of the population, and A weighted by one. This means that the relatedness is -1/(N-1) as Hamilton (1971) first showed.

again this has a negligible effect when the population is large. I have presented the geometric view of relatedness rather informally in the hope of being intelligible. The two connections of the geometric view, with social behaviour and kinship, can be made more rigorously, and this is the task of the next two sections. By making the connections in a more mathematical way, we prepare for answering the question of how generally Hamilton's rule applies, the topic of section 6.

4.   Hamilton's rule and regression

The connection of the geometric view of relatedness proposed in the previous section on the one hand, with the definition of r from section 2, and with the notion of common ancestry, on the other, must be made within some formal framework. Regression is the framework I shall use. It is ideally suited because it has an algebraic side, in which to do things, and a geometric side, in which to think about what to do and how to do it. Relatedness was first treated as a regression coefficient by Hamilton (1970) and later by many others. However, in all cases they use only the algebraic aspect of a regression, as a covariance divided by a variance. The interpretation of these regressions in geometrical terms is the basis of the view of relatedness expounded in the previous section.

The first step is to find an algebraic way of saying that the average position of R, which is the average position of the potential recipients of A, lies on the line from μ to A. If it is a fraction β of the way from μ to A, then we can write

$$\mathbf{E}(R) = (1-\beta)\mu + \beta A,$$

or, after slight rearrangement,

$$\mathbf{E}(R) = \mu + \beta(A - \mu).$$

The average value of R depends on the position of A and, if we imagine that A can vary, formally that A is a random variable, then we should write the average value of R conditional on the value of A as the left-hand side of this, giving

$$\mathbf{E}(R|A) = \mu + \beta(A-\mu). \tag{8}$$

This is in statistical terms a regression equation, that is, it gives the expectation of one variable conditional on the value of another. It is a linear regression equation. We can compare this to the more familiar form, often used to represent the model in a simple, univariate regression of y on x in statistical textbooks:

$$\mathbf{E}(y|x) = \alpha + \beta x,$$

or

$$y = \alpha + \beta x + \text{error}$$

The differences between these two regressions are as follows. In the first, the dependent variable is R, and the independent variable is $A - \mu$, while in the second the dependent variable is y and the independent variable is x. Those are merely notational differences. Usually, α is unknown and needs to be estimated, while we regard μ in eqn (8) as known. Finally, μ, A, and R may be vectors in some large dimensioned space [the 'genic structures' of Jacquard (1974)], whereas α, x, and y are ordinary numbers. For ease of explanation, I want to discuss the simple case where μ, A, and R are ordinary numbers, and represent a p-score of the population, actor, and potential recipient, respectively. (The case where they must be treated as vectors is important when we want to estimate relatedness from electrophoretic data).

We have now expressed in an algebraic way the statement that R lies at a certain fraction of the way along the line from μ to A. The next step is to find a connection between this algebraic statement, and the algebraic definition of r in section 2. To do this, we will pretend we are trying to estimate β from our regression eqn (8), using the set of occasions in one generation as data. Of course, we don't have that data as a set of numbers in any example, but we can nevertheless find an algebraic expression for what we would estimate β to be if we did.

What form does this data take? Recall the notation of section 2. On the jth occasion, the actor's p-score was D_j and the potential recipient's p-score was R_j. There were J occasions in all. So the (x, y) data points we have with which to estimate β are the J pairs $(D_j, R_j - \mu)$. In order to see how to estimate β, it is convenient to rewrite eqn (8) in a more convenient form, as

$$R_j = \mu + \beta(D_j - \mu) + \epsilon_j, \tag{9}$$

where ϵ_j is the deviation of R_j from its expected value.

We could rearrange eqn (9) to give

$$\beta = \frac{R_j - \mu}{D_j - \mu} - \frac{\epsilon_j}{D_j - \mu}$$

and this suggests taking the first term on the right-hand side as an estimate of β. The reasons are that we know all the terms in it, whereas ϵ_j is unknown, and that the second term is on average 0 because ϵ_j is. However, this would give us a separate estimate of β for each occasion and, as our biological hypothesis implies that the set of occasions can be taken together, we want a combined estimate from all J occasions. Accordingly, we multiply eqn (9) by arbitrarily chosen constants z_j giving

$$z_j R_j = z_j \mu + \beta z_j (D_j - \mu) + z_j \epsilon_j,$$

and then adding up over all occasions to get

$$\Sigma z_j R_j = \Sigma z_j \mu + \beta \Sigma z_j (D_j - \mu) + \Sigma z_j \epsilon_j$$

and rearranging as before to give

$$\beta = \frac{\Sigma z_j (R_j - \mu)}{\Sigma z_j (D_j - \mu)} - \frac{\Sigma z_j \epsilon_j}{\Sigma z_j (D_j - \mu)}.$$

This suggests in the same way as before an estimate b of the form

$$b = \frac{\Sigma z_j (R_j - \mu)}{\Sigma z_j (D_j - \mu)}. \tag{10}$$

This is an unbiassed estimate of β provided only that the z_j are not correlated with the ϵ_j. Now I introduced the z_j as arbitrary constants and promised to return to them. A standard statistical argument goes on from eqn (10) to find the most efficient choice for these arbitrary constants, from the point of view of minimizing the sampling variance of b while maintaining its property of unbiassedness. However, we part company with the statistical argument here and choose to compare eqn (10) with our definition of r in section 2.

Let me repeat the definition, which was eqn (7), and set it alongside eqn (10).

$$b = \frac{\Sigma z_j (R_j - \mu)}{\Sigma z_j (D_j - \mu)}, \qquad r = \frac{\Sigma H_j (R_j - p)}{\Sigma H_j (D_j - p)}.$$

These two equations are very similar. One notational difference is that the population mean p-score is represented by μ on the left and by p on the right. Apart from that, we can make the two right-hand sides identical by choosing our arbitrary constants z_j to be equal to H_j.

The identity of these two formulae means that our algebraic definition of r from section 2 is an estimate of a regression coefficient. Further, it is an estimate of the regression coefficient that says how far R lies along the line from μ to A. Hence, our geometric picture of relatedness is the same as our algebraic formula, which was derived so as to make Hamilton's rule work. This is the formal version of the purely verbal argument made in section 3 that the geometric view of relatedness was also right for making Hamilton's rule work.

The main work of the section is now completed, but there is one important matter to discuss. We saw that r can be regarded as a special case of b, an estimate of the regression coefficient that expresses how far R lies along the line from μ to A. However, what if R does not lie on that line? In section 3 we assumed that it did, and now we return to look again at that assumption. The next section considers under what circumstances we expect the assumption to be fulfilled: here I want to explain the meaning

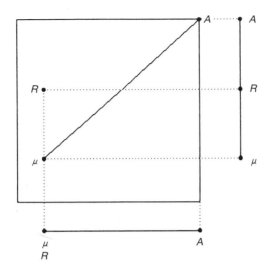

Fig. 6. When R lies off the μ–A line, the relatedness is different for different loci. In this example, the relatedness at the locus represented on the vertical axis is 0.5, while the relatedness at the locus represented on the horizontal axis is 0.

and consequences of its being false. The meaning is clear enough, and is illustrated in Fig. 6. R falls to one side of the line that joins μ and A. The immediate consequence is also illustrated in Fig. 6. We can compute r for any p-score. If we choose first the gene frequency at the locus represented on the horizontal axis, then we find that $r=0$, while if we choose the p-score to be the gene frequency at the other locus, then $r=0.5$. So, if R lies off the μ–A line, then r between the same sets of individuals is different for different p-scores, and therefore for different alleles and different loci.

One consequence of this is extremely important, and it is that we cannot expect to apply Hamilton's rule simply and usefully. Hamilton's rule is most simply applied when we know a group of individuals and some part of their genealogies. We then expect closer kin to be more co-operative and altruistic towards each other. This is because we hope to know r from common ancestry. However, if R lies off the μ–A line, then the r that makes Hamilton's rule work varies from locus to locus, and p-score to p-score, and so cannot depend on ancestry alone. The p-scores that are of special interest are those that control social behaviour, but it is beyond our current abilities to discover which loci are involved in those p-scores, to assess the influence of each allele at each locus, and to measure the presence of each allele in each individual. Even if we could do this, much of the value of Hamilton's rule would be lost. It is the rule's simplicity and range of applicability that make it so useful.

Another consequence is subtler. In the formula for b, our estimate of a regression coefficient, we have a set of arbitrarily chosen constants, namely the z_j. Consider to what extent the value of b depends on which set of arbitrary constants we choose. If R lies on the μ–A line, then the value of b

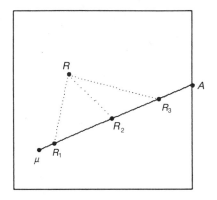

Fig. 7. The arbitrary constants determine the slope of the projection of R onto the μ–A line which is used in the computation of relatedness. If R lies on the μ–A line, then a projection with any slope will simply leave R where it is. However, when R lies off the μ–A line, as in the diagram, different projections take R to different points on the line, leading to different relatednesses.

does not depend at all on which constants we choose. If it lies close to the line, then the value of b depends weakly on the choice of arbitrary constants; and if R lies far from the μ–A line, then b will depend strongly on the choice. The reason for this is illustrated in Fig. 7. The regression involves projecting the position of R onto the μ–A line, and the choice of arbitrary constants determines the slope of the lines used in the projection.

Our definition of r from section 2 is, as we have seen, equivalent to a choice that the arbitrary constants should be chosen to be H_j. The H_j are the phenotypes of the actor on the jth occasion: H_j equals 0 if the actor does not perform the act and 1 if it does. It is the only item in the definition of r that contains information about dominance or about the way in which the p-score is related to phenotype at all. It follows that if R lies on the μ–A line, then dominance can have no effect on the direction of change of a p-score from one generation to the next. However, if R lies off the μ–A line, then dominance can affect the direction of the change. Queller (1984a) discusses a kin selection model where dominance has an effect and I will discuss briefly how this can be interpreted in terms of the geometric view of relatedness.

The example involves triploid endosperm in plants. This is a tissue that acts as an intermediary in the flow of resources between the parent plant and a seed. In the simplest of the genetic arrangements, the endosperm receives the same genetic contributions from each parent as the seed does, but it receives the maternal contribution in double dose. The endosperm is therefore triploid. The kin selection question is how the endosperm will be selected to act in its crucial role. Will it have the same interests as the parent, or as the seed, or somewhere in between?

This problem is the subject of a number of papers and this paper is not one of them. I want to concentrate on only one aspect of it and that is what

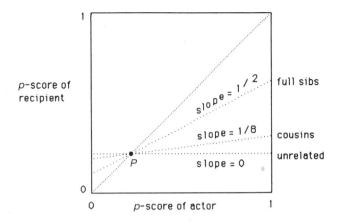

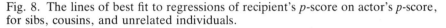

Fig. 8. The lines of best fit to regressions of recipient's *p*-score on actor's *p*-score, for sibs, cousins, and unrelated individuals.

is the relatedness of endosperm to seed? Let us use the simplest kind of *p*-score, the frequency of one allele. Figure 8 is helpful here and shows another side to the geometric view of relatedness. It has the actor's *p*-score on the *x*-axis, and the potential recipient's *p*-score on the *y*-axis. The large point marked *P* is on the 45° line, and is the point corresponding to the population mean on both axes. Imagine each occasion plotted on this graph, using the *p*-scores of the actor and the potential recipient, and calculating the line of best fit, on the condition that the line goes through the point *P*. (This is the same as redrawing the figure with *P* as the origin, and calculating the best fitting line that goes through the origin.) The slope of this line is the relatedness. Figure 8 shows lines of best fit (to imaginary data) for diploid siblings, cousins, and unrelated individuals. This graph is the basis of a method of estimating relatedness from electrophoretic data (Pamilo and Crozier 1982). Figure 8 is similar to Fig. 1 of Orlove (1975).

The statement that *R* lies on the μ–*A* line can be interpreted in terms of Fig. 9. For diploids, we can summarize the data that would appear in Fig. 8 as three points, each with its own weight: the mean *p*-score of potential recipients when the actor's *p*-score is 0, the mean *p*-score of the potential recipient when the actor's *p*-score is 0.5, and the mean *p*-score of the potential recipient when the actor's *p*-score is 1. The weights are the number of occasions contributing to each point. Now there always is a line of best fit, whether these three points are in a straight line or not. If *R* lies on the μ–*A* line, as in Fig. 9a, then these three points do lie on a straight line that passes through the point *P*. The consequence of this is that no matter how we weight the different points, the best fitting straight line is the same.

If, on the other hand, the three points do not lie on the same line, as in Fig. 9b, then how we weight them will affect which is the best fitting straight line and so its slope, and so will affect *r* according to our definition of section 2. Two factors determine the weights given to these three points.

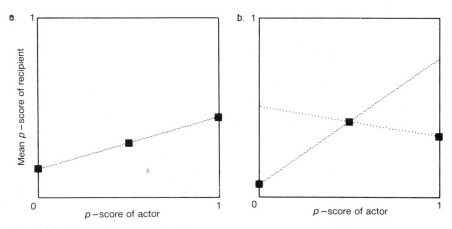

Fig. 9.(a) The points occupied by squares represent the average p-score of recipients when the actor's p-score is 0, 0.5, and 1. When they lie in a straight line, that line is the best fitting straight line. Therefore, the slope of the best fitting straight line does not depend on the weights given to the points in the regression. (b) Here the three squares do not lie on a straight line, and the slope of the best fitting straight line varies between the two lines illustrated. In this case, the slope of the best fitting straight line does depend on the weights given to the three points in the regression.

The number of occasions on which the actor has each of the three genotypes is one factor, and the other is the weights H_j which enter into the definition of r.

The case of the triploid endosperm is illustrated in Fig. 10. If the endosperm is homozygous, then the seed is also homozygous. If the endosperm is heterozygous, then the seed is too. However, both kinds of heterozygosity in the endosperm (one-third and two-thirds of an allele) are

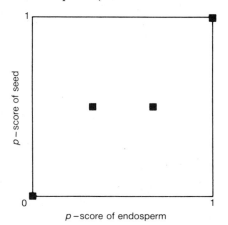

Fig. 10. A plot of the genotype of the seed (vertical axis) against the genotype of the endosperm (horizontal axis). The points do not lie on a straight line, so the best fitting straight line depends on how the points are weighted in the regression.

associated with the same kind of heterozygosity in the seed (one-half of the allele). Hence, the points do not fall on a straight line. Suppose that the frequency of an allele determines whether the endosperm transfers extra resources from the parent to the seed. Then H_j equals 1 if the endosperm does transfer extra resources, and equals 0 if it does not. H_j is a weight in finding the best fitting line and so the best fitting line depends on the dominance rule. This helps us to understand why, as Queller (1984a) and Bulmer (1985) show, whether such an allele spreads or not depends on its dominance. I am very grateful to Dr M. G. Bulmer for drawing my attention to this example.

The first lesson from this example is that it is important to know when R lies on the μ–A line. When it doesn't, as in the case of the triploid endosperm, it is not possible to predict the direction of selection of a character without knowing its genetics. This makes it much harder to see how natural selection will have acted on a character of whose genetics we are ignorant (which is to say nearly all characters). The second lesson is that our example is rather peculiar, and we may hope that this kind of difficulty is uncommon generally. We shall see that this example falls within the category of inbreeding, which turns out to be an important and general exception. The commonness of this difficulty is the topic of the next section.

5. When does R fall on the μ–A line?

To answer this question, we need a universe of possibilities. This will be provided by the mathematical machinery of identity by descent (Malécot 1969; Jacquard 1974). Within this set of possibilities, we will distinguish between those cases where R does and those in which R does not fall on the μ–A line. As we saw in the previous section, this is an important point in the usefulness of Hamilton's rule. To the extent that R does fall on the line, r between the same sets of individuals is the same for all loci and all alleles.

The plan of the section is to introduce the concept of identity by descent, to explain how this concept is used to describe in a precise way the genetic consequences of shared ancestry, and then to set out the assumptions needed to establish that for any non-inbred pattern of shared ancestry between actor and potential recipient, R will fall on the μ–A line. The plausibility of these assumptions in various circumstances will be discussed. We will then turn to the way in which genetic similarity between members of the same local population can be dealt with. The reason why inbreeding causes difficulties will be treated briefly. Then, as we have been using the machinery of identity by descent, we must confront the paradox of inbreeding of Seger (1981). The whole point of this section is concerned to avoid the conclusion that r varies between alleles and loci, and so at the end I explain briefly what would be the evolutionary consequences if it did.

Falconer (1981) defines identity by descent relative to a base population which existed at some time in the past. Two genes are said to be identical by descent if they are both descended from the same gene in that base population. If we allow the base population to be two generations ago, then we can calculate how many genes at one locus are shared identically by descent by two outbred diploid sibs. There is a 50 per cent chance that they receive a copy of the same gene copy from their father. Even if their father is a homozygote at that locus, the two alleles he passes to his offspring are not identical by descent if one is a copy of his paternal allele and the other is a copy of his maternal allele. There is an independent 50 per cent chance that they receive a copy of the same gene copy from their mother. We can therefore assign an equal probability of 25 per cent to each of four possibilities: the sibs share no genes identically by descent at that locus, they share only the paternal gene, only the maternal gene, or both.

The genetic consequences of genealogical relationships are summarized by numbers of this sort. For non-inbred diploid relationships, a set of numbers called Cotterman's coefficients are used. They are the probabilities that the two individuals share exactly none, exactly one, or exactly two alleles identically by descent. For possibly inbred diploid relationships, a set of eight numbers is needed if the distinction between the individuals' maternal and paternal alleles is not kept. The details of these fascinating mathematical constructions can best be pursued by consulting Jacquard (1974), but only two points are essential to us here. They are that any specified genealogical relationship has a representation of this sort, that can be written down, and that one quantity of particular interest can be computed from this representation. That quantity is what Crozier (1970) called G, and is the average fraction of alleles in one individual (in our case the recipient) that are identical by descent with any of the alleles in the other (the actor). A proviso that will be important is that the probabilities which characterize a genealogical relationship are correct only in the absence of selection.

Let us denote by φ the average fraction of the potential recipient's genotype that is identical by descent with any of the alleles in the actor. Now in principle, φ could be different for different actors' genotypes, and this is indeed the case in relationships that involve inbreeding, as we shall see later. For the present, let us confine ourselves to non-inbred relationships and assume that φ is the same for all genotypes. We can think of two separate parts of the recipient's genotype, the IBD part and the non-IBD part, and in particular of the gene frequencies of those fractions. How do they compare with the population mean gene frequency and with the actor's gene frequency? Our route will be to assume that the IBD part of the recipient's genotype has on average the same gene frequency as the actor does and that the non-IBD part has on average the population mean gene frequency. Let us look at the consequences of these assumptions and then ask when we might expect them to be true.

If both these assumptions are true, then we can write the average gene frequency of the recipient as

$$\mathbf{E}(R \mid A) = \varphi \text{ (average gene frequency of IBD part of genotype)}$$
$$+ (1 - \varphi) \text{ (average gene frequency of non-IBD part of genotype)}$$
$$= \varphi A + (1 - \varphi)\mu$$

$$\mathbf{E}(R \mid A) \quad = \mu + \varphi(A - \mu) \tag{11}$$

This equation is similar to one of Jacquard (1974, p. 118) with notational differences and it is also familiar to us from above. It is the same as eqn (8) eqn (8) of section 4 except that here φ has taken the place of β. Furthermore, φ is determined by the genealogical relationship of the two individuals, and so is the same for all alleles and for all loci. Here we find a justification for assuming that R lies on the μ–A line. Equation (11) is true for any outbred genealogical relationship and also for mixtures of outbred genealogical relationships. If half of the potential recipients were sibs, a quarter were cousins and a quarter were unrelated, then the probabilistic statements of gene identities could be made for this mixture, and φ calculated in the same way as for 'pure' relatives.

Hence, it seems to be a general conclusion that if the genetic connection between the actor and the recipient arises through common ancestry with no inbreeding then R will lie on the μ–A line, but we must not forget the two assumptions we made to arrive at this conclusion. The first was that the IBD part of the recipient's genotype had the same gene frequencies as the actor did, and the second was that the non-IBD part of the recipient's genotype had the same gene frequencies as the population mean. Nor must we forget the proviso that the calculation of φ from the pattern of common ancestry assumes the absence of selection. Let us take the two assumptions in turn, and see why they should be true, and then decompose them into more fundamental assumptions. The justification of these more fundamental assumptions and of the proviso about the calculation of φ will be the task of the next section.

Why should the IBD part of the recipient's genotype have the same gene frequencies as the actor? If the actor is a homozygote at a locus, then the IBD part must be identical unless a mutation has occurred in the path through the genealogical tree that connects the recipient and the actor. After all, if it is identical by descent, then it is surely identical. If the actor is a heterozygote, then the recipient's IBD part must be identical (again, barring mutations) to one of the actor's alleles, but to which? If we can assume that it is equally likely to be to any of the actor's alleles, then it would again follow that the recipient's IBD part was the same on average, though not necessarily the same in any particular case, as the actor. We can sum up the assumptions needed to justify this in the statement that the only force at work is the random segregation of Mendelian genetics. Specifically, we need to assume no mutation and no selection.

Why does selection disturb our conclusion that the IBD part of the recipient's genotype is on average the same as the actor's genotype? Imagine tracing back along a diploid genealogical tree from the actor to a common ancestor and then forward again to the recipient, computing at each stage the chance that an allele in the actor was present. The rule we use is that there is a 50 per cent chance that a gene in a parent is passed on to an offspring, and that there is a 50 per cent chance that a gene in an offspring came from each parent. However, we know that each individual in the path survived to reproduce and if survival to reproduce is different for different genotypes, then this means that certain alleles had more than a 50 per cent chance of being passed on, while others had less. The presence of selection means that Mendel's rules are not enough. Furthermore, the effect of selection will be different for different alleles and for different loci.

Noting that the first assumption has decomposed into the more fundamental assumptions of no mutation and no selection, we turn to the second assumption, which is that the non-IBD part of the recipient's genotype has the same gene frequencies as the population mean. This means that from the actor's point of view, the non-IBD part of the recipient's genotype is of equal importance to the genotype of a randomly chosen member of the population. This can fail in two ways. It may be that some genotypes are particularly prone to find themselves in the role of recipient. Alternatively, it may be that the population is not homogeneous, and that different parts of the population have different mean gene frequencies. For heterogeneous populations it may be that the non-IBD part of the recipient's genotype has the same gene frequencies as the local mean, not the global population mean. The actor is also likely to be genetically more similar to the local population, and so this is an additional source of genetic similarity between the actor and recipient, besides their common ancestry. Let us then add two more to our list of more fundamental assumptions: that the tendency to be a recipient is not affected by genotype and that the population is homogeneous.

While the strict truth of eqn (11) depends on no mutation and no selection, it will be approximately true provided the mutation or selection is small. What does small mean? The error in eqn (11) that is caused by a given mutation or selection pressure depends on the length of the genealogical routes connecting the recipient and the actor, because the opportunity for mutation and selection to act is proportional to the length of those paths. Only very strong selection could materially alter the conclusion that an offspring is equally likely to have either of its mother's alleles. Suppose, on the other hand, that an individual is known to have one gene identical by descent with a heterozygote ancestor a thousand generations ago. The accumulation of a thousand generations' worth of selection could easily make it much more likely that one rather than the other of the ancestor's genes is the one that is shared. Incidentally, we can also see how selection can affect the calculation of probabilities of identity by descent. Of all the individuals' ancestors a thousand generation's ago, it is more likely that the individual shares genes at a locus with those ancestors that possessed the selectively advantageous genes rather than the selectively disadvantageous ones. Which ancestors those are will be different for different loci.

Hence, both for the assumption that the IBD part of the recipient's gene frequencies is the same on average as the actor's gene frequencies, and for the calculation of probabilities of identity by descent, the mutation rates and selection pressures must be small in comparison with the lengths of the genealogical paths that connect the actor and the recipient. The justification for making this assumption about selection and mutation, in an argument that is intended to show how selection works on social behaviour, is one of the purposes of the next section.

Now the other fundamental assumption, of homogeneity, is also important and I want to consider two different ways of dealing with a structured population in the framework of the chapter. The first way is to assume that the non-IBD part of a recipient's gene frequencies is the same as the local population mean. Formally, we can let μ_L represent the mean of the local population to which the interactants belong and rewrite eqn (11) as

$$E(R \mid A) = \mu_L + \varphi(A - \mu_L) \tag{12}$$

Now eqn (12) is no longer of the same form as eqn (8), and so we cannot identify φ and r so easily. The vital difference is that φ_L is no longer the same for every actor A. The derivation of r was valid for a population subdivided in any way: it does not change in parallel with the change in φ. Hence, this way of dealing with a subdivided population is analytically possible, but would imply a more complicated connection between r and φ, involving the correlation between A and μ_L. It would be different for different alleles and different loci, and so would lose the most attractive features of the simpler model.

The approach taken by Hamilton (1975) is equivalent to the introduction of a regression that says how μ_L varies with A, as follows:

$$E(\mu_L \mid A) = \mu + \delta(A - \mu),$$

leading to

$$E(R \mid A) = \mu + \{\delta + (1 - \delta)\varphi\}(A - \mu).$$

The problem with this is that δ, the regression coefficient that says how diverse groups are, may well differ from allele to allele and locus to locus, for the same reasons as we shall encounter below that impede the second way by which we might try to deal with local groups.

This second way is to view the extra similarity between group members as arising because of extra kinship ties between them. If a group is small, then random mating within the group must often be incestuous within the wider context of the population as a whole. Mating between relatives is one cause of extra genetic similarity between group members (another is adaptation to local conditions). Can we then not deal with social interactions between group members by calculating φ taking these extra kinship ties into account? The extra kinship ties are no doubt hard to specify in any particular case, but might this allow us in principle at least to rescue the identity between r and φ in the case of local groups?

Unfortunately not. The reason is that we fall foul of the fundamental assumption of no selection in a serious way. The paths that make fellow group members genetically more similar to each other than they are to members of other groups are likely to be long. The reason that selection influences the paths of descent is that it changes gene frequencies. It follows that random drift can have the same effect. Although random drift does not influence these paths averaged over all possible present universes, we are interested in predicting selection in this particular present universe. Therefore, to apply in a simple way the machinery of identity by descent to model the effects of local grouping would be triple folly: the paths are long, significant drift and selection are very likely to have occurred, and inbreeding is almost certainly involved. An additional complication is that to the extent that groups are genetically isolated, the members of a group are likely to be competing particularly with each other for genetic representation in future generations. To that extent, the relevant relatedness will be relative to the local group and not to the whole population. I have discussed this at more length elsewhere (Grafen 1984). The application of the geometric view here is therefore rather complex, and I will pursue it no further here. The case of subdivided populations is nonetheless an important problem.

Now we tackle the problem of inbreeding, namely why φ should be different for different genotypes, that is the recipient lies at a different fraction of way from μ to A for different positions of A. Figure 11 illustrates the possibility.

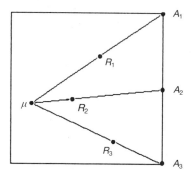

Fig. 11. Under inbreeding, the relatedness may depend on the genotype of the actor. Here the homozygotes have a higher relatedness than the heterozygote has to their respective recipients. It is likely in these circumstances that an actor will be differently related to the recipient at different loci.

Each actor represented there has a different value of φ, and if we considered more than one locus of one actor, we would find that it would have different values of φ at different loci. (As we saw in section 3, this implies that in the genetic space representing all the loci, R would not fall on the μ-A line.) The reason can be seen in Fig. 12a, which shows the various patterns of identity by descent associated with outbred sibship, and Fig. 12b, which shows the patterns for sibship when the parents were themselves outbred sibs. Each pattern has a value of φ associated with it, which is the fraction of the recipient's genotype that is identical by descent with any part of the actor's genotype at that locus. Each pattern also has a probability of occurring in a given genealogical relationship. The average value for the sibship is some weighted average of these φs. Now the three patterns for outbred sibship make no connections between the actor's two alleles and so place no restrictions on the actor's genotype. It follows, conversely, that knowing the actor's genotype tells us nothing to make any one of the three patterns more likely than another.

Some of the inbred patterns, on the other hand, imply that the actor's two alleles are identical by descent and therefore that the actor must be a homozygote. Arguing backwards, if we know that the actor is a heterozygote, it follows that those patterns of identity by descent are not possible. Therefore, to find the average φ for an individual known to be a heterozygote, we must average over the patterns in which the actor's alleles are not identical by descent. A homozygote may or may not have its two alleles identical by descent, and so any of the patterns may occur. However, the knowledge that the actor is a homozygote makes it more likely that they are identical by descent. The balance of likelihoods depends on the frequency of the allele the homozygote possesses and so the relatedness will be different for the different types of homozygote. At a locus with three alleles, for example, there will therefore be four different degrees of relatedness: one for all heterozygotes and one each for the three different homozygotes (Elston and Lange 1976).

The difference between homozygotes and heterozygotes is important because dominant alleles are expressed in heterozygotes, while recessive alleles are expressed only in homozygotes. The spread of a recessive allele will therefore be determined by the homozygotic relatedness, while the spread of a dominant allele will be determined by some average of the homozygotic and heterozygotic relatednesses. The dominance of an allele may therefore affect whether it spreads or not. The relatedness of homozygotes depends on allele frequency, so whether an allele spreads or not may depend on its frequency. All these complications mean that alleles with the same phenotypic effect may be selected in opposite directions because they differ in dominance or frequency. Michod (1979) and Michod and Anderson (1979) have particularly clear discussions of this effect of inbreeding. The difficulties in applying Hamilton's rule to cases with inbreeding is discussed by Michod (1982), and a number of examples are worked through by Uyenoyama (1984). The appropriate definition of relatedness in terms of the nine identity coefficients of Fig. 12b, gene frequency and dominance, is given by Michod and Hamilton (1980, formula 3).

Our next topic is the paradox of inbreeding (Seger 1981). In the introduction, we dismissed the concept of identity by descent because we saw that if we look back far enough, most gene identity will be identity by descent. Yet for the whole of this section, we have been using identity by descent quite freely, refusing to see the paradox by looking no further back in time than some base population. This device serves the purposes of population geneticists well and involves no trickery, for they have a base population with respect to which they wish to measure identity by descent (Falconer 1981). We, on the other hand, have no such natural base population.

Our 'natural base' is not a population at all, but rather the behavioural rules that the population can adopt. If a nestling behaves in a particular way towards others in the same nest, but cannot discriminate between

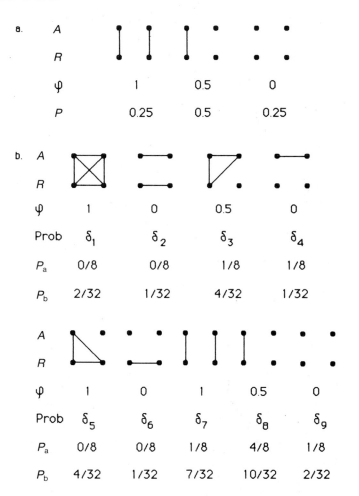

Fig. 12.(a) The top dots in each square represent the genotype of the actor, and the bottom two represent the genotype of the recipient. A line connects each pair of alleles that are identical by descent. Depicted here are the three patterns that are possible between outbred sibs and P is the probability with which they arise. φ is the fraction of alleles in the recipient that are identical by descent with any allele in the actor. With outbreeding, each genotype is distributed between the patterns in the same proportions. This allows us to calculate the relatedness between actor and recipient by averaging the φs weighted by the P's. (b) The nine possible patterns of identity by descent between diploids are illustrated, following Jacquard (1974). The δs represent the probability with which each pattern occurs and are his condensed identity coefficients ('condensed' because no distinction is made between an individual's maternal and paternal alleles). Two examples of the δs are given (P_A: the actor's two parents are parent and sibling of the recipient; P_B: the actor and recipient are sibs whose parents are outbred sibs). The top row contains those patterns in which the actor's two alleles are identical by descent, that is to say, the actor is inbred. The probability of this is therefore $\delta_1+\delta_2+\delta_3+\delta_4$. The bottom row contains the patterns in which the actor is not inbred. φ is the fraction of genes in

them in its actions, then the category of 'fellow nestling' is a natural one. For any particular nestling and fellow nestling, there will be a complex history of gene identities, but we shall divide it into two parts. One part will be correlated, in the sense that fellow nestlings are more likely to share it with each other than they are with nestlings from another nest; while the other part is uncorrelated, because fellow nestlings are no more likely to share it with each other than they are with those from another nest.

Take the simplest case, where all nestlings are full sibs, and there is random mating in a finite population. The correlated part of the relationship between nestlings is only sibship, because random mating destroys all other correlations. It does this in two steps. The random mating last generation prevents correlation between the alleles of one parent (that is, inbreeding of the parents), and the random mating this generation prevents any correlation arising from common ancestry between the parents, taking the set of sibships in the population as a whole. It may be that in a particular case, the mates are sibs, but this occasional event with a high positive correlation between the mates will be balanced by the much commoner slight negative correlation between the genotypes of mates guaranteed not to be sibs.

Of course, if nestlings can sense whether their parents are sibs or not, and can behave differently accordingly, then the systematic part of the relationship becomes more complex, but the same principles apply. The fact that parents are sibs, for example, would be a correlated part of the category 'Fellow nestling in a nest of mated sibs'. Any relationship that is uncorrelated with the distinctions that nestlings can make in their behaviour will cancel out when averaged over the population.

The resolution of the paradox of inbreeding is that relatedness is a property of 'action categories', and this has interesting implications. It does not make much sense to compute from their common ancestry the relatedness between two particular individuals. They will be related in very many ways, by distant routes, and the conclusion from adding them all together would probably be that the relatedness was one (Jacquard 1974, p. 171)! It makes more sense to ask how categories of individuals are related, such as nestlings or playmates or locals. Their relatedness can be assessed from common ancestry by knowing the paths that are correlated with the distinctions that individuals can make in their behaviour. It is

---

the recipient that are identical by descent with any allele in the actor. The average value of φ for inbred actors is therefore $(\delta_1+\delta_3/2)/(\delta_1+\delta_2+\delta_3+\delta_4)$. The average value for outbred actors is $(\delta_5+\delta_7+\delta_8/2)/(\delta_5+\delta_6+\delta_7+\delta_8+\delta_9)$. Heterozygotes cannot be inbred, so are restricted to the bottom row, and therefore have the outbred relatedness. Homozygotes may be inbred or outbred, with a probability that depends on the frequency of the allele they bear. All the standard measures of relatedness can be defined in terms of the δs. Inbreeding coefficients for actor and recipient are $f_A = \delta_1+\delta_2+\delta_3+\delta_4$, and $f_R = \delta_1+\delta_2+\delta_5+\delta_6$. Malecot's 'coefficient de parente' is $f_{AR} = \delta_1+(\delta_3+\delta_5+\delta_7)/2+\delta_8/4$. Wright's coefficient of relationship is $r_A = 2f_{AR}/\{(1+f_A)(1+f_R)\}^{1/2}$. Hamilton's regression coefficient of relatedness is $b_{RA} = 2f_{AR}/(1+f_A)$.

important to stress that the centrality of those distinctions is quite natural. Our purpose in using relatedness is to analyse social behaviour, whose evolution must depend on the powers of discrimination animals possess. Jacquard (1974, p. 171) stated that 'It is obvious that, from this point of view, an inbreeding coefficient cannot be regarded as an estimate of a real quantity, but is simply a measure of information', and relatedness would fall into the same category as inbreeding. From our point of view, relatedness is a measure of the animal's information, and in the evolution of social behaviour this is something very real and very important.

Another consequence is that one individual can have a different relatedness to another, depending on the 'action-category' to which it belongs at the time. Suppose in a species with much brood parasitism that nestlings are related to each other by a quarter, because some are full sibs while others are unrelated; and that nestlings recognize each other only as fellow occupiers of the same nest and so cannot recognize each other after leaving the nest. Then an individual will treat sibs and unrelated individuals alike when in the nest, with a relatedness of one-quarter, and later will treat all of them as unrelated because he cannot distinguish them from any other individuals. Thus, relatedness is a property of 'action categories', not of individuals and not simply of patterns of common ancestry.

Finally, in this section whose main purpose is to argue that R will often fall on the μ–A line, I turn briefly to the evolutionary consequences of R's lying off that line. Recall that this means the relatedness as defined by eqn (7) of section 2, our definition of r that makes Hamilton's rule work, will be different for different p-scores. [It was Hamilton (1967) who first discussed this problem of genomic discord, with reference to the difference between the autosomes and the sex chromosomes, a cause I have entirely neglected here. He explored the consequences for the sex ratio of the fact that the X chromosome follows the same pattern of relatedness as haplodiploids, while the Y chromosome has an extreme pattern in which a male has a relatedness of one to sons and father.] Our formula for $w \triangle p$, eqn (6), is still correct for any p-score, so let us consider how it affects the frequency of an allele by choosing the p-score to be the frequency of that allele. In particular, let us consider those alleles that influence the performance of the social action. Then there is a critical value of r, say r', defined by $r' = c/b$, which is the relatedness at which the allele's frequency does not alter as a result of selection on the social action. The benefit and cost are such that with a relatedness of r', the actor is indifferent towards performing the act.

Now, if r varies between alleles and between loci, then alleles that contribute towards performance will increase in frequency if their r is greater than r' and decrease if it is less. Alleles that reduce performance of the act will increase in frequency if their r is less than r', and decrease if it is greater. Thus, selection will be acting in opposite ways on alleles with the same effect. The net effect of this genomic discord would depend on the relative numbers and size of effect of the alleles with different values of relatedness. Alleles with extreme values would tend to go to fixation or extinction. Some kind of average would prevail, but selection could stop only when all alleles affecting the social behaviour, that had not gone to fixation or extinction, had the same relatedness. The position is more complicated than this, because relatedness is not a property of an allele by itself. If relatedness varies through the genome, then it is likely that the relatedness for one allele depends on its frequency and on the frequency of other alleles. This variation is described by eqn (7) in section 2, where the complication arises because the weights H_j matter and depend on these genetic complications. The conclusion that R falls on the μ–A line is useful because r does not then depend on those weights.

The next section picks up the fundamental assumptions on which we base our conclusion that R falls on the μ–A line and justifies them for our purposes. This forms part of a defence of Hamilton's rule as a general evolutionary principle.

6. Hamilton's rule as a general evolutionary principle

The justification of Hamilton's rule as an evolutionary principle has two parts to it. The first part was begun in previous sections and is to be completed in this. It is that we expect characters to evolve under conditions in which Hamilton's rule will be obeyed, and so if we understand a character correctly in nature, we should observe Hamilton's rule being obeyed. We justify the first part by showing that Hamilton's rule is a correct summary of a certain particularly relevant set of population genetics models. While the population geneticist is rightly interested in the exact analysis of a wide range of models, some models will be more relevant than others to the likely effect of selection in nature. The first part is permissive, it says that if we do use the rule we should get the right answer. The second part, not mentioned so far, is an argument that says we should want to analyse a character in terms of Hamilton's rule. For both models and data, this will enable us to understand the evolution of a character in a particularly interesting way; and for certain kinds of data it may be possible to analyse a character in terms of Hamilton's rule when it is not possible by rival methods. The second part of the justification may sometimes be so strong that it is worth using Hamilton's rule even when it is an incorrect summary of relevant population genetics models. These assertions will be argued later on.

Let us make sure of the starting point for our current argument by reviewing the relevant conclusions of earlier sections. On the way, we will come across points that need further discussion. The fundamental result is eqn (6) of section 2, which shows that Hamilton's rule correctly describes the effect of social interactions on the direction of selection of any p-score, provided we define relatedness as the genetic regression coefficient described in eqn (7). We must also remember that eqn (6) is based on an additive model of social interactions. The usefulness of eqn (6) depends on the relatedness being the same for all p-scores. Then, in section 4, we saw that this condition is fulfilled if the genetic similarity between interactants arises through links of common ancestry that do not involve inbreeding, provided there is weak selection and the population is homogeneous.

This leads us to the conclusion that Hamilton's rule, using the relatedness we would compute from ancestry, works under the conditions of (i) additivity, (ii) weak selection, (iii) homogeneity, and (iv) outbreeding. As we shall see in the next section, these conclusions are not new, and are the consensus of a fair-sized literature. If the third or fourth of these assumptions is broken, then alleles with the same effect may be selected in opposite directions because they differ in frequency or dominance. Although circumstances in which these assumptions are broken are very important, I have nothing further to say here about the outcomes of this genomic discord.

For most of this section, I shall concentrate on recent common ancestry as the main cause of the genetic similarity that influences the evolution of social behaviour. As a partial defence, I now consider a very special property of common ancestry in this respect. Common ancestry produces genetic similarity at every locus in the genome, and it produces the same genetic similarity (as measured by the relatedness of section 2) at every locus. Population structure as a cause of genetic similarity has already been discussed. Other possible mechanisms that have been suggested are genetic determination of micro-habitat choice, 'green beards', and the active detection of genetically similar individuals. Now I want to picture the effect

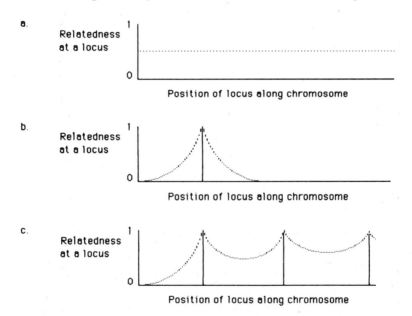

Fig. 13.(a) The relatedness between outbred diploid sibs under conditions of weak selection is the same at all loci and for all alleles at each locus. (b) When genetic similarity is caused by ensuring that one locus is identical, the relatedness falls off on either side of the 'guaranteed' locus. The rate of decay depends on the linkage disequilibrium between that locus and nearby loci. (c) To maintain a high relatedness along the chromosome, it is necessary to have 'guaranteed' loci at close intervals, so that all loci are in linkage disequilibrium with one or more of the 'guaranteed' loci, which can then act as telegraph poles.

of these possible mechanisms, and to consider their likely evolutionary consequences; I will not discuss the difficult problem of whether the mechanism is likely to evolve in the first place.

In the picture, each chromosome is a line, and the relatedness at each locus is plotted above it. Figure 13a shows the simple case of interacting sibs, where the relatedness at each locus is a half. Figure 13b shows the relatedness caused by any mechanism that ensures genetic identity at one particular locus. The relatedness at that locus is 1, and the relatedness at closely linked loci is increased above 0. The other loci must be so closely linked that they are in linkage disequilibrium with the locus that is guaranteed identical. The likely evolutionary consequence of one guaranteed identical locus is small. If a character could be affected only by very close loci, then it might evolve under a Hamiltonian regime with a relatedness appreciably different from 0. If a character can be affected by loci that are distant from the one guaranteed identical locus, then Hamilton's rule will continue to operate with 0 relatedness at distant loci and the net effect of selection will probably be settled in favour of the more numerous distant loci. This argument is the one given by Hamilton (1967) to explain the comparative inactivity of the sex chromosomes. To build up the kind of substantial average relatedness across the genome which is likely to be needed to produce an evolutionary effect, the mechanism would have to ensure identity at a number of loci, to be the 'telegraph poles' of Fig. 13c.

Hamilton (1964, p. 25) first suggested the possibility of direct recognition of possession of certain alleles [later termed 'green beard genes' by Dawkins (1976)] or even traits, and it has been further discussed by Dawkins (1976, 1982) and by Rushton et al. (1984). There are two possible effects of a direct recognition mechanism of this sort. The first is to distinguish between different kinds of relatives. Thus, in a mixed nest of sibs and half-sibs, the more similar individuals are likely to be sibs. This could be useful information. The other effect is to recognize from among unrelated individuals a subset who are as genetically similar as, say, cousins, but who are in fact genealogically unrelated. It is to this second possible effect that I now turn.

Rather than ask how a mechanism could ensure identity at enough loci to hold up the 'telegraph wire' of relatedness for the whole genome, I want to ask a logically prior question. It is, what fraction of unrelated individuals happen to be as genetically similar to an individual as its cousins? In other words, forget for the moment about how these individuals are to be recognized, how many of them are there to be discovered in the first place? If the number is very small, then genetic recognition mechanisms are probably unimportant for detecting genetically similar organisms from the population at large, but may still be important for distinguishing between different kinds of relative. We can answer this question in a rough way by recalling how many unrelated individuals we have met who are as phenotypically similar to us as our sibs (cousins, second cousins, and so on) are. We can also answer the question in a mathematical way, by calculating the probability distribution of relatedness [using eqn (7) from section 2 and weighting each locus equally] among unrelated individuals. Suppose there are L units in the genome that we can assume are in linkage equilibrium, and that the mean probability of gene identity at a locus is f. Then, using the normal approximation to the binomial, we can obtain a normal deviate corresponding to a given relatedness. It is defined by the following formula:

$$ z = r \sqrt{\left\{ L \ \frac{1-f}{f} \right\}} $$

For a species with $L=50\ 000$ and $f=0.8$ (plausible values for humans; Cavalli-Sforza and Bodmer 1971; Nevo 1978; Lewontin 1974), the square root is about 100. Hence, the normal deviate corresponding to an identical twin arising by chance among unrelated individuals is 100, for sibs is 50, and for cousins is about 12. These are very high normal deviates. A standard deviate of 4.8, corresponding to a relatedness of 0.048, would make the frequency of such individuals one in a million. A standard deviate of 3.1, corresponding to a relatedness of 0.031, would make their frequency 1 in 1000. If these values of L and f are reasonable, then it seems that there are no great advantages to be gained. The advantage from interacting with one individual with a relatedness of 0.03 would have to be offset against the cost of testing all 1000 individuals. I conclude that for genetic recognition to work other than by distinguishing between different kinds of relative, it is necessary that the relevant part of the genome be small, and this brings us back to green beards and the necessity for linkage; or alternatively that genetic variability be low. Unless the genetically similar, but unrelated individuals are of comparable frequency with corresponding relatives, selection is unlikely to favour complex or costly mechanisms for their detection. Of course, it may be that in some species these computations work out much more favourably.

I proceed by concentrating on common ancestry as the major cause of genetic similarity that is relevant to the selection of social behaviour. In the previous section, we saw that under weak selection, additivity, homogeneity, and outbreeding, common ancestry leads to unanimity of relatedness among the alleles and loci, and that this relatedness is derivable from knowledge of the ancestral links between interactants. I now wish to discuss how the assumptions of weak selection and additivity can be defended. The test is whether the r defined by eqn (7), a measure of genetic similarity I will call Hamilton's r because it makes Hamilton's rule work, is different from the relatedness that would be calculated from common ancestry, which I shall call the ancestral r.

An important point to begin with is that we are mostly interested in the evolution of a character, as distinct from the genetic changes that take place at a particular locus. The major part of the defence of the assumptions of weak selection and additivity is that they are likely to hold when selection has brought a character to a state in which there are no large improvements to be made. The argument is one given by Fisher (1958). Although strong selection pressures are to be expected when a character is changing rapidly, perhaps because of some change in the environment or in another species, once most of the required change has been made the possible improvements are small. It follows that the only strong selection pressures in connection with the character are downwards pressures on strongly disadvantageous mutants. When the 'fine-tuning' of the character takes place, the only relevant selection pressures are weak ones. From now on we will assume where necessary that selection is weak. This is reasonable because we are interested in the conditions under which characters are perfected by natural selection. In this respect, Hamilton's rule and the Darwinian principle that animals are designed to maximize their reproduction, are in the same position. The Darwinian principle depends on fine-tuning under conditions of weak selection so that evolution will produce the precise optimum. Hamilton's rule depends on fine-tuning additionally because the weak selection implies equality between Hamilton's r and the ancestral r. Relying on the same condition twice introduces little extra burden of assumption for Hamilton's rule.

The simplest kind of character is one which can take any value in a continuum, and for which the fitness of the character is a smoothly varying, single peaked function of its value. (In the case of social interactions, the fitness function may depend on the state of the population in some way but this complication is unimportant.) I shall call this kind of character graded, and first discuss how Hamilton's rule works for them. Then we will go on to consider apparently ungraded characters.

In section 2 we proved Hamilton's rule using a model with additive fitness interactions and by agreeing to define r in a special way. We now wish to show that Hamilton's rule applies using ancestral r and with any kind of fitness interaction, on the assumption of weak selection. We saw in the previous section that the assumption of weak selection guarantees that Hamilton's r is the same as the ancestral r, and this leaves us with the additivity of fitness interactions. The meaning of additivity involves our model of the fitness of an individual as a function of its phenotype and of the phenotype of the individuals with which it interacts. Equation (4) from section 1 expresses the model as

$$w_i = \frac{1}{l_i} (f_i + n_i y_i b - m_i h_i c) \tag{4}$$

where w_i is the fitness per haploid set of individual i, l_i is its ploidy, f_i is its baseline fitness, n_i is the number of interactions in which it is the potential recipient, y_i is the average phenotype of the actors on the occasions when individual i is the potential recipient, m_i is the number of occasions on which it is the actor, and h_i is its phenotype. (For more details on the meaning of these symbols, the reader is referred back to section 2.) In our present application we are interested in whether a slight variant in a form of behaviour in a social interaction will spread or not. b is therefore the average effect on the recipient of interacting with the variant form rather than with the common form, and c is the average effect on the actor of adopting the variant form of behaviour rather than the common form. Examples of variants are to give slightly more or slightly less food, or to be slightly more or slightly less vigilant. b and c must be small if the social interaction is close to evolutionary stability, and the variant action is advantageous.

The important part of additivity is that w_i should be approximately a linear function of h_i and y_i. Alternative specifications for w_i are discussed by Scudo and Ghiselin (1975), Cavalli-Sforza and Feldman (1978), Charlesworth (1978), Maynard Smith (1980), Feldman and Cavalli-Sforza (1981), and others. The meaning of additivity is that the effects of separate occasions on an individual's fitness add up. Being a recipient once increases an individual's fitness by a certain amount. It is possible that being a recipient again does not increase the individual's fitness again by that same amount. This may be because, for example, the first gift of food rescued the individual from starvation, whereas the second merely allowed it to put on a little extra fat. Help in the form of warning calls when a predator is near does not add, but has diminishing returns, because the chance of survival is increased more by the first than by later calls.

However, when selection is weak the effects of occasions will approximately add up. This follows if fitness is a continuous and smooth function of the phenotypes of the interactants, which is part of the definition of a graded character. This convergence to additivity under weak selection is not a new point and has been made many times. One implication of this is that b and c may vary for a given kind of social interaction as the frequencies of those adopting it change [Uyenoyama and Feldman (1982) make this point more generally].

To find if Hamilton's rule fits a social interaction well, it is therefore necessary to consider the effects of slight variations in the behaviour of interactants. The total benefit and cost of the interaction are relevant to the question of whether the actor should abandon the interaction altogether, but marginal benefits and costs are relevant to the question of whether Hamilton's rule fits the precise form of the interaction.

Now this 'marginalization' of Hamilton's rule works well for graded characters, in which continuous changes in behaviour affect fitness continuously and there is only one local optimum. However, many characters seem to be not at all graded, and it is to these that we now turn. There are two main ways in which characters may fail to be graded. The first is that an action may be 'all or nothing', as in Haldane's famous example of a man saving a child from drowning by diving in and pulling him from a river (Haldane 1955). It is convenient here to suppose the victim to be an adult. The second way is that there may be 'multiple peaks' in the adaptive landscape (Wright 1977, 1978), and so it may be important to know whether a mutation of large effect will spread. When we need to predict the behaviour of a mutation of large effect, we lose the assumption of weak selection, and so lose the useful conclusion that Hamilton's r and ancestral r are equal.

My main strategy here is not to tackle ungraded characters directly, but rather to argue that apparently ungraded characters may be graded after all. This will not be a compelling case that all apparently ungraded characters are in fact graded, because that is largely a matter of fact. Instead I aim only to suggest that many apparently ungraded characters may be graded. After making this case, I will turn briefly to the question of how ungraded characters might evolve.

In Haldane's example, the action of the hero who dives in to save someone from drowning seems to be 'all or nothing'. The problem this causes for Hamilton's rule is as follows. The potential hero makes an imaginary computation of the gene frequencies of the victim and, in particular, of the frequency of an allele that has a strong influence on the tendency of its bearer to be a hero in those circumstances when the opportunity arises. The obvious computation for the potential hero to make is the one from section 3. The gene frequency of the victim is in part the same as his own and the rest is the same as the population mean. That part is the relatedness.

The complication is that the potential hero has extra information about the genotype of the victim. The victim is still alive. This suggests that if the victim has had an opportunity to save others from drowning, he has not taken it. (We suppose a hero risks his own life.) This in turn suggests that the victim does not have alleles that are conducive to saving others. This information means that Hamilton's r at the loci affecting the tendency to save others from drowning is lower than their genealogical relationship alone suggests. If opportunities are rare, then this extra information is weak and the ancestral r will be only a little lower than Hamilton's r. If they are common, then the extra information would be stronger.

This illustrates the important point that genealogical relatedness affects the evolution of social behaviour only through the tendency of the same alleles to be present in actor and recipient, and so any other information about the presence or absence of those alleles is relevant in exactly the same way. We saw earlier that animals are unlikely to have information that makes a non-relative as genetically similar as a relative at all loci in the genome. The possibility shown in Haldane's example is that because of the phenotypic effects of particular loci, mere survival can give information about the genotype at those loci. In this case, the loci are those that affect willingness to risk one's own life. Selection through kin effects will be altered at those loci, in a way that Hamilton's rule predicts. Although Hamilton's rule is correct in this case, it loses its most appealing feature, because Hamilton's r is not the same as ancestral r.

Now the definition of a graded character was that the fitness of the character is a smoothly varying, single peaked function of its value. My aim now is to show how the character 'reaction when faced with the opportunity to dive in to a river to save someone else's life at some risk to one's own' might be graded, despite the all or none aspect of the decision in a particular case. These opportunities for heroism will not all be the same. In some cases the chance of success will be high, in some cases low. The risk will be great on some occasions and small on others. Various clues will be available to these chances, such as the temperature of the water, the light, the distance from shore, the swimming ability of the victim, the presence of other potential rescuers, and so on. The character can be graded once we accept that the behaviour of the potential hero depends on these circumstances. For any given set of circumstances, his decision is discrete – he dives or he doesn't – but to represent his decision rule we must say that in such and such sets of circumstances he will jump, and in so and so circumstances he will not.

Let us consider how those decision rules may differ among individuals. It

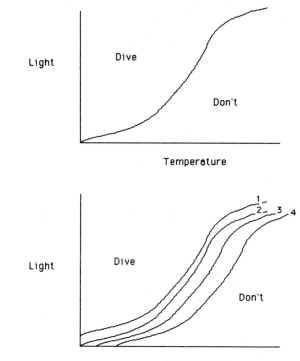

Fig. 14.(a) An imaginary decision rule that divides possible circumstances into two sets – those in which Haldane's man would dive and those in which he wouldn't. (b) The imaginary decision rules of four individuals. If the circumstances in which the rules differ are rare, then there is little information about genotype to be gained from observing the consequences of having dived or not (such as survival in the example in the text), even though diving may be common.

helps to have a picture and so Fig. 14a illustrates an imaginary decision rule that depends on temperature and light. The area to the left and above the line represents combinations of temperature and light at which he will dive, while the area to the right and below represents combinations at which he will not dive. Figure 14b illustrates the decision rules of a number of different individuals and I have assumed that they differ only slightly. For the sake of argument, assume that when an opportunity arises, it is equally likely to occur at any combination of circumstances in the figure. How does this affect the calculation of gene frequencies made by a potential hero?

The answer is that it tends to restore them to ancestral values. If individuals in the population have similar decision rules, then occasions on which they would disagree are rare. Most of the occasions will therefore fall into the unanimous regions of Fig. 14b, and so the survival to date of the victim gives very little information about the position of his decision line. His survival therefore also gives very little information about his genotype at the loci affecting his decision line. It follows that Hamilton's r and the ancestral r will be much the same, because ancestry has again become the only guide to genetic similarity. If any shape of decision line is attainable by selection, then this means that the position of the line will be determined under a Hamiltonian regime and so at the circumstances on the line $rb-c=0$, while above it where the decision is to dive, $rb-c>0$, and below it where the decision is not to dive, $rb-c<0$. This is a strongly Hamiltonian conclusion, suggesting that in each circumstance the decision to dive or not will be determined by Hamilton's rule using ancestral r. This is possible because in the final evolved state of the character there is no genetic variation in the behaviour displayed in most circumstances. This is a good example of the divergence between the population geneticist who is interested in the genetics of a character, and the ethologist who is interested in the likely outcome of its evolution.

The crucial question for the equality of Hamilton's r and ancestral r is whether further information on genotype is available to the actor. There will be little information if the opportunities are rare on which the decisions of different individuals would be different.

This process of converting an all or nothing response into a graded character might be called parametrization, and is a generalization of the use to which Charlesworth (1978) put penetrance. The penetrance of a gene is the probability with which it is expressed, but usually there is no suggestion that the circumstances in which it is expressed are any different from the circumstances in which it is not. If the decision to dive in Haldane's example were made according to an irrelevant cue, then this would be equivalent to a random decision.

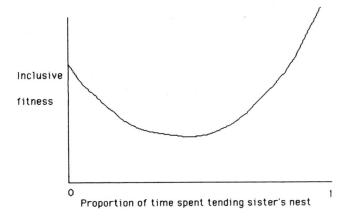

Fig. 15. A possible relationship between inclusive fitness (using the ancestral value of relatedness) and the extent of helping a sister. When a mutation for complete help arises in a population that doesn't help at all, the relatedness at that locus towards a sister is not the same as the ancestral relatedness, because the fact that the sister is not helping a previous sister makes it likely she does not share the gene for helping.

The second reason why a character may fail to be graded is that there are multiple peaks in its adaptive landscape (Wright 1977, 1978). An example of this is shown in Fig. 15. If the population is in the 'catchment area' of the highest peak, then our previous argument for graded characters applies, and the fine-tuning of the character is performed under Hamiltonian conditions in which Hamilton's r equals ancestral r. If the population is in the catchment area of a lower peak, then so far as small changes are concerned, the same argument applies again. The potential problem for Hamilton's rule occurs when a mutation occurs that could take the population from one catchment area to another. This could be an advantageous mutation of large effect, for which Hamilton's r and ancestral r might be different. If so, a mutation that is advantageous from the point of view of the ancestral r, and so from the point of view of a student of their behaviour, could fail to spread because Hamilton's r was lower than the ancestral r.

A hypothetical example of this, again borrowed from Charlesworth (1978), involves a hymenopteran female faced with a choice of how much of her time to allocate to her own reproduction, and how much to allocate to the reproduction of her sister, who, we may suppose, has nested nearby. It is convenient to suppose that a succession of sisters come to nest in an area. The value of $rb-c$ plotted against the fraction of time spent helping her sister might well be U-shaped, as in Fig. 15, on the grounds that dividing her time between two different nests would be inefficient. If the population is all at the lower, selfish peak, then a mutation of small effect would not spread. Consider the fate of a dominant mutation of large effect that caused a female to abandon her own attempt to reproduce and instead to become a full-time assistant to her sister. The fact that her sister has not abandoned her own reproduction and become an assistant to an earlier sister, shows that she does not share this mutation of large effect. Consequently, when this mutation of large effect arises, Hamilton's r will be 0 for this allele, despite the fact that the ancestral r is 0.75. (If the mutant had been recessive, Hamilton's r would have been 0.5. Notice how, when Hamilton's r and the ancestral r differ, genetic details such as dominance affect Hamilton's r.)

This character can be graded in the same way as Haldane's diving. If the mutation is of low penetrance, or if its expression is conditional on particularly favourable circumstances, then the fact that the sister has not helped another sister is much less informative about her genotype, and Hamilton's r will be close to ancestral r.

The conclusion is that the two most obvious reasons why a character may seem not to be graded are not decisive. Characters that have an all or nothing aspect and characters that have multiple peaks in their adaptive landscapes may after all be graded. If so Hamilton's r and the ancestral r will be equal when the character is finely-tuned by selection, and so the form of the character will be as predicted by Hamilton's rule using the ancestral r.

Even when characters are ungraded, fairly extreme circumstances must obtain to make the two r's very different. The fact of being a potential recipient must be very informative about an individual's genotype to separate the r's, and this has strong implications for the size of the phenotypic effect, for the informative nature of its expression, and for the genetics of the character. It is for these reasons that I expect Hamilton's r and ancestral r to be the same for most characters. This is merely one opinion about an unknown matter of fact, but it is unfortunately important for the view we hold of the usefulness of Hamilton's rule.

We have reached the end of the exposition of the first part of the justification of Hamilton's rule. The rule is a reasonable summary of a particularly important set of population genetics models, namely those that seem most likely to be relevant to the outcome of the evolution of a character. Thus, if we use Hamilton's rule, we should not get the wrong answer about the evolutionary value of a character. If the first part says that it is safe, the second part of the justification says that it is desirable to analyse a character in terms of Hamilton's rule. Let us take models and data in turn.

To analyse a character in terms of Hamilton's rule means to work out r, b, and c. Now in some models, in particular those devoted to testing whether or not Hamilton's rule works, r, b, and c are taken as given, and in that case there is no analysing to be done. There are other models in which the character is described or defined in some other way, and it is in these models that Hamilton's rule can be used to advantage. The questions of particular interest for which this analysis is useful include: does this character spread because of its effect on the actor's own number of offspring, or does it spread only because of the effect of the action on the number of offspring of others? Is the action altruistic, or selfish, or spiteful? The way a character is defined may obscure these points. For example, the effects of an action may be described as, first, an effect on self and, second, an effect on every group member. b is the net effect of the action on self, and so if self is one of the group members, we must subtract the second effect from the first to compute b. Similarly, c is the total effect on others, and so to obtain c we must multiply the second effect by the number of group members (besides self) receiving it. This particular conversion from one description to another reconciles the trait group selection of Cohen and Eshel (1976), Matessi and Jayakar (1976), and Wilson (1975), with the Hamilton's rule approach.

Note that b and c are differences in number of offspring. Part of the discrepancy between the additive and multiplicative models pointed out by Cavalli-Sforza and Feldman (1978) is simply one of measurement. A physicist who decides to measure mass in a logarithmic scale can easily refute Newton's second law, if he fails to make the necessary concomitant adjustment to its algebraic expression. He deserves no attention because Newton's second law is framed with particular scales of measurement in mind. Hamilton's rule is designed to work on differences. It is not surprising that if they are measured as ratios instead, then Hamilton's rule fails. If the differences caused by the social action vary from occasion to occasion, then it will be necessary to find the average differences to compute b and c [Uyenoyama and Feldman (1982) explore this more rigorously]. However, simply measuring the costs and benefits as differences will not remove the discrepancies between the multiplicative and additive models if some individuals partake twice in social actions, because then there is an added complication. The number of social actions an individual is involved in may depend on its genotype, because an altruist is more likely to have altruistic relatives. If, in addition, the effects of social actions combine multiplicatively, then the effect of the marginal social action when measured as a difference will depend in a systematic way on genotype. Thus, the benefit received by a recipient may depend on its genotype and this causes problems for inclusive fitness theory (Seger 1981; Queller 1984b). The main point here, though, is that even when social actions are thought to act multiplicatively, it is still possible and desirable to measure the costs and benefits as differences for the purposes of Hamilton's rule, averaging over occasions where necessary.

Having seen how and why we might want to analyse models in terms of Hamilton's rule, let us now turn to data. The questions Hamilton's rule will help us answer are the same – is the action selfish, spiteful, or altruistic, and is the action advantageous only because of the fitness effect on others? In applications to data, Hamilton's rule comes into its own. The great differences from models are that usually with data on social traits, the genotypes of individuals are unknown and the genetic system controlling the trait is unknown. This makes worries about dominance, number of loci, and mode of gene action purely academic. In modelling, the fundamental population genetics method of finding the number of offspring of each genotype is the main rival to Hamilton's rule. This alternative simply cannot be applied to data if the genotypes of individuals are unknown. Hamilton's rule can be applied, provided enough information is available to measure the effects of the social action. If this information is not available, we cannot discover by any means whether the action is altruistic or not. I have worked out an example in detail elsewhere (Grafen 1984, section 3.3.3), using data of Noonan (1981) on joint nesting in *Polistes fuscatus*.

Hamilton's rule, then, is the way to answer central questions of interest about a social action. It has the advantage that, unlike other methods, it can be applied to data that is available about social actions. For these reasons, the theoretical investigations of Hamilton's rule are unlikely to replace it with an alternative so far as data is concerned. A method that works well in a model in which individuals' genotypes are known can fail altogether when they are unknown. The theoretical investigations are valuable in finding the scope of validity of the rule, those circumstances in which it correctly predicts the direction of evolution of a trait. However, in a case where it is known to be incorrect, say where inbreeding is present, the practical response is to apply Hamilton's rule anyway, and treat the results with caution.

In conclusion, Hamilton's rule is useful because it tells us whether an action is selfish, spiteful, or altruistic, and because it tells us the value we expect one individual to place on another's reproduction. These are its important points, and they apply equally to models and to data. The current interest in altruism and social evolution has the rule at its centre, for it embodies the definitions of selfish, altruistic, and spiteful proposed by Hamilton (1964).

Many genetic details turned out not to upset the validity of Hamilton's rule. Dominance and the exact form of fitness interactions do not matter in a range of cases and even apparently discontinuous characters may be continuous underneath. However, there are limits in the range of possible genetic mechanisms for which Hamilton's rule does predict exactly the direction of evolution. The most important limitations are that when the population is structured, or when there is inbreeding, the relatedness which will make Hamilton's rule work will vary from allele to allele, and from locus to locus; so alleles with the same phenotypic effect will be selected in opposite directions because of differences in dominance or frequency. The likely outcome of these complications is not known, but it seems probable that some kind of average relatedness will prevail. The required modification to Hamilton's rule is likely to be a procedure for computing that average relatedness from knowledge of the population structure, nature of the inbreeding, and possibly the frequency of mutations of particular kinds.

The viewpoint of a population geneticist on the exceptions was expressed by Uyenoyama and Feldman (1982). They say:

'In summary, for additive kin selection models our analyses indicate that Hamilton's theory is remarkable precise' (p. 616).

'We show that by regarding the multiplicative model as an additive model with genotype-dependent benefit parameters, the multiplicative model can be reconciled with Hamilton's theory' (p. 626).

They do not think that the multiplicative model should be regarded as a minor variation on the additive model, however, and they give two reasons. The first is that multiplicative models allow strong internal equilibria, that is strongly stable polymorphisms, and the second is that the nature and identity of internal equilibria are affected by the use of approximations. These are good reasons why population genetic theorists should be interested in multiplicative models, but they are not reasons why an ethologist should be. The theory tells the ethologist that the behaviour of organisms should follow Hamilton's rule, and that is the only part of the theory he is likely to be able to test. Whether the population is genetically uniform for this behaviour (Uyenoyama and Feldman's 'viability-analogous equilibrium'), or polymorphic (their 'structural equilibria'), is less important, and will usually be beyond his ability to find out.

There are exceptions, in population genetics models, to Hamilton's rule. The lesson I draw from them is that, in order to make the rule work in those circumstances most relevant to the outcome of evolution on a character, we should find a suitable generalization of relatedness. The rule relates facts observable in the field to the evolutionary fate of a character, and its terms are a touchstone to the evolutionary significance of social behaviour. It is for these reasons, and not merely because it is a summary of certain population genetics models (though it is surprisingly good at that too), that Hamilton's rule is a general evolutionary principle.

7.   A brief survey

In this section, a brief survey is presented of the work on Hamilton's rule and relatedness, and the ideas presented in previous sections will be placed in that context. For a full review of derivations of Hamilton's rule, see Michod (1982). The two authoritative sources for definitions of relatedness are Michod and Hamilton (1980), and Seger (1981). The first theme of this section is the confirmation of Hamilton's rule in a variety of models. The second theme is a division of claimed exceptions into three categories. The third is a comparison of the definition of r in this chapter with previous definitions.

Hamilton (1964) presented the derivation of his rule with care, and left the reader in no doubt about the qualifications and complications in the analysis. It was an additive model and so would work only for small effects; r was a good guide to the likely genotypic constitution of relatives only under weak selection. Since then, the rule has been rederived and these qualifications restated many times. Michod (1982) reviews these rederivations; here I wish only to mention the categories into which they fall. Some models have a single locus, while in others the tendency to altruism is a quantitative or polygenic character. Some ('inferential') models infer the gene frequency in recipients from the gene frequency of donors and the relatedness; while in others ('grouped') the population is divided up into mutually exclusive groups. These two classifications divide possible derivations into four, but to my knowledge only three of the possible types exist.

The original derivation was single locus and inferential, as were those of Hamilton (1970), Charnov (1977), Orlove and Wood (1978), Michod (1979), Harpending (1979), Charlesworth (1980), Charlesworth and Charnov (1980), Seger (1981), and the expository derivation of Maynard Smith (1982). The single locus grouped models include Hamilton (1975), Orlove (1975), Levitt (1975), Scudo and Ghiselin (1975), Charlesworth (1978), Cavalli-Sforza and Feldman (1978), Boorman and Levitt (1980), Uyenoyama and Feldman (1981, 1982), Uyenoyama et al. (1981), O'Donald (1982), Wade (1982), and Karlin and Matessi (1983). I know of no inferential quantitative models, but grouped quantitative models have been given by Yokoyama and Felsenstein (1978), Boyd and Richerson (1980), Aoki (1982), Crow and Aoki (1982), Engels (1983), and Cheverud (1984, 1985). All these models confirm the rule under the condition of weak selection, in the absence of inbreeding and additional population structure.

The derivations of Hamilton in 1964 and especially in 1970 have an extra generality, in allowing any number of interactions of any kind between the members of the population. Most others, including the derivation in this paper, consider only a single kind of action, involving only two individuals at a time, that may in some derivations be repeated. The 1970 derivation is in many ways still the most comprehensive of all, as it is also valid for inbreeding (though not mentioning exactly the problems this may cause). Almost all of the present paper can be regarded as a long, expository footnote to Hamilton's 1970 article.

The three kinds of approach have merits and disadvantages. The connections between the inferential and grouped models are discussed by Abugov and Michod (1981), and Michod (1982), who call them 'inclusive fitness' and 'family selection' models. The 'inferential' models are more general because the grouped models require that the social interactions take place within groups, and that within groups no discrimination is made between further and nearer relatives. The grouped models, on the other hand, allow greater rigour as a consequence of this simplicity.

The hallmark of a rigorous derivation, besides algebraic rectitude, is complete recursion. This means that when our 'generation-to-generation' equations are applied to one generation, they tell us enough about the next generation to allow us to apply to our generation-to-generation equations to that next generation. If they do, we can then move on to the third generation in the same way, and on to the fourth, fifth, and so on. Our derivation in section 2 is an example of an incomplete recursion, because eqn (6) needs to know about covariances in this generation, but tells us only about the mean p-score in the next. It follows that to obtain a complete recursion, we would need to find a way of calculating what those covariances will be in the next generation. Another name for completeness of recursion is dynamic sufficiency (Lewontin 1974).

Complete recursions are very demanding. Little vagueness is allowed, so it would be difficult, if not impossible, to prove a result that held for all different kinds of relatives. Each separate case must be analysed on its own. The joy of Price's method is its generality, and it is potentially a great advantage, but this raises an important question. Naively, it seems that either complete recursions are necessary to avoid error, in which case Price's advantage is illusory, or complete recursions are unnecessary. No doubt both methods have their part to play, but so far none of the important conceptual advances have been made by completely recursive methods. The important advances I have in mind are the original derivation (Hamilton 1964); the application of covariance methods and the 'backwards' definition of r as 'whatever will make Hamilton's rule work' (Hamilton 1970); the derivation of the rule to include inbreeding and the evaluation of relatedness from common ancestry in the presence of inbreeding (Hamilton 1970); the derivation of the rule for grouped populations (Hamilton 1975); and the rederivation that showed more explicitly the problems inbreeding may cause for Hamilton's rule (Michod 1979). The completely recursive methods tend to follow on in the rear, providing the comfort of a more rigorous derivation some time after the advance party has decided the problem and found the solution. When the incomplete methods lead to a result as simple as eqn (7), then for those interested in the fairly gross behaviour of the system it may be that there is little extra that complete recursions can do. Complete recursions are necessary to determine the exact nature of interior equilibria when the variations in r caused by selection are enough to change the sign of $rb-c$. These computations are unlikely to have any observational, empirical significance. Of course, it may be that complete recursions will play an important role in future in resolving the outstanding problems of inbreeding and heterogeneous populations.

The quantitative and polygenic models are all grouped models. They perform (essentially) an analysis of variance on the character of 'tendency to perform altruism to group members', and on its heritable component. There is, then, a within group variance and a between group variance in each. These imply certain correlations between group members and can be used to compute the covariances in terms of which relatedness can be defined (Michod and Hamilton 1980). The different models show, to varyingly explicit degrees, that Hamilton's rule applies under conditions of weak selection to indiscriminate social actions within groups in a grouped population.

Engels (1983), in addition, explicitly modelled what, in section 6, we argued by words, that a succession of small mutations will take a population to the equilibrium determined by Hamilton's rule. Cheverud (1984) developed quantitative genetics models that include maternal effects; he confirmed Hamilton's rule in the case with no maternal effects. Uyenoyama (1984) analysed exact models with specific patterns of inbreeding. She stated that the relatedness calculated from ancestry was a good guide to the models' results even when the rule strictly failed. The available literature confirms that the rule is a remarkably good summary of a wide range of population genetics models.

The derivation of section 2 is an inferential model, but is both single locus and polygenic, according to the interpretation of the p-score. It allows arbitrary ploidies, individuals may interact more than once, and the interactants need not have the same gene frequencies as the population. It thereby combines in one derivation results that would otherwise have to be proved separately.

The chief purpose of many of the papers cited above in support of Hamilton's rule was to prove it wrong in particular cases. The second theme of this brief review is a classification of the alleged exceptions to the rule. I divide them into three categories. The first are the strong selection exceptions that arise through breaking the weak selection assumption of section 6. These are real exceptions, but in section 6 I argued that they were likely to be unimportant from the point of view of the outcome of selection on a character. The second category are those that arise through a misinterpretation of Hamilton's rule, and these are unfair exceptions that we can set aside. The third category are exceptions within the assumptions of weak selection, which we may call important exceptions.

The multiplicative models begun by Scudo and Ghiselin (1975), and popularized by Cavalli-Sforza and Feldman (1978) are in danger of falling into the second, 'unfair' category because, in Hamilton's rule, the effects of an action are expressed as differences not ratios. This is well discussed by Uyenoyama and Feldman (1982). They regard the multiplicative model as an additive model with genotype dependent cost and benefit parameters, and this allows the results of their model to be compared with Hamilton's rule fairly. They found in their exact model of sib interaction that Hamilton's rule worked under weak, but not strong selection, and so we may place their exceptions, and by extension those of previous multiplicative models as well, in the first category.

Another strong selection exception is the deviation depending on heritability found by Boyd and Richerson (1980) in a quantitative model.

I will mention three examples of unfair exceptions. The first is claimed by Uyenoyama *et al.* (1981), in an additive model with two sexes. They define the benefits and costs relative to a mean fitness of unity, separately within each sex. The relative fitness of an unhelped, unhelping male is therefore 1 and that of a helped male is $1+\beta$. However, when all males are helped, the mean 'relative fitness' of the males is $1+\beta$, and so their fitnesses must be scaled down because the total male fitness is fixed by the number of females and the sex ratio of their offspring. Thus, in a population where virtually all males are helped, a rare selfish mutation reduces the male's number of offspring not by β, but by β normalized to the mean male relative fitness, i.e., $\beta/(1+\beta)$. This, together with a similar adjustment to the costs, brings the models' results into complete agreement with Hamilton's rule.

The second example of an unfair exception is from Cheverud (1984). He proposes that there is pleiotropy between two characters, namely some trait expressed as a juvenile and a maternal effect on that trait. He finds that Hamilton's rule, with costs and benefits derived from the effects of the juvenile trait alone, does not predict the effect of selection on the trait. A pleiotropic gene has two effects and it is hardly surprising that we cannot predict its fate with a rule applied to only one of those effects. It is certainly not a shortcoming in the rule.

The third example of an unfair exception is from Queller (1984b). He presents a two strategy, two player game theory model of interactions between relatives, where the strategy adopted by the recipient can affect the pay-offs to both interactants. This dependence of benefits and costs on the recipient's genotype is just the sort of thing that will plausibly cause exceptions to Hamilton's rule. In one of his two models (the continuous case), Queller supposes that the two strategies are played with a probability that depends on genotype, and he searches for an ESS probability which once common cannot be invaded by any other probability. Using a subscript 'Q' to distinguish Queller's notation where necessary, the pay-offs to strategy one are $b_Q - c_Q + d_Q$ when playing against strategy one, and $-c_Q$ when playing against strategy two. The pay-offs to strategy two are b_Q when playing against strategy one, and 0 when playing against strategy two. The idea is that strategy one is to perform a social act that costs c_Q, and has benefit b_Q, but that when both interactants perform this social act, there is a non-additive interaction so that instead of each receiving $b_Q - c_Q$, they each receive $b_Q - c_Q + d_Q$. He finds that the condition for a population playing strategy one with probability P to be invaded by a slightly higher probability of playing strategy one is

$$rb_Q - c_Q + Pd_Q + rPd_Q > 0$$

where r is the relatedness between interactants. This seems to differ from Hamilton's rule by the presence of the third and fourth terms which are caused by the interaction d_Q. However, the b of Hamilton's rule is the average effect on the opponent's pay-off (in game theory terms), and this depends on the probability with which the opponent plays the two strategies. Similarly, c is the average effect on self and, in Queller's model, this also depends on the action of the opponent. Working out those average effects, we find that:

$$b = b_Q + Pd_Q$$

and

$$c = c_Q - Pd_Q.$$

When we express the condition for the spread of a slightly higher probability of strategy one in terms of b and c, we recover Hamilton's rule, for

$$rb_Q - c_Q + Pd_Q + rPd_Q = rb - c.$$

An equivalent conversion from game theory notation to Hamilton's rule notation was performed by Grafen (1979). The recovery of Hamilton's rule in this case is encouraging, because it may often be the case that the recipient's genotype affects the benefits and costs of social actions.

These unfair exceptions show that care must be taken in applying Hamilton's rule. The rule is based on a particular way of measuring costs and benefits, and on a particular concept of relatedness. If the rule works with one set of interpretations of its elements, then it is most unlikely to work with another. A result has no interest *as an exception to Hamilton's rule* if it is based on the wrong interpretation of r, b and c.

Now we come to the important exceptions, those where even with the correct interpretation of the rule and with weak selection the rule fails. The only example I am aware of is inbreeding and the reasons were first given in a slightly different context by Seger (1976). He was interested in explaining inbreeding depression as adaptive altruism by homozygotes, who by reason of their homozygosity were more highly related than average to those around them. The problems posed by inbreeding for Hamilton's rule were first explicitly discussed by Michod (1979), and Michod and Anderson (1979). The problem, as discussed in section 6, is that homozygotes and heterozygotes are differently related towards the same classes of relatives. This makes dominance important. The relatednesses depend on gene frequency, which adds to the complication.

Faced with this problem, a first approach is to try to place bounds on the deviation from the simple rule. This can be done using formula (3) of Michod and Hamilton (1980), although they did not do so. The relatedness in this context we may think of as the critical cost-benefit ratio at which the allele's frequency will not change. In one example of Michod (1979, in Fig. 2b with $h = 1$), the relatedness changed from seven-sixteenths to four-sixteenths as the gene changed in frequency from 0 to 1. (Michod and Hamilton pointed out that owing to a mistake in an earlier formula, Michod's figure is in error in showing the relatedness becoming negative as the gene frequency approached 1.) This is an appreciable difference, of a size which could at some time matter in an application of Hamilton's rule to date.

One line of attack is to find bounds for a particularly relevant subset of inbred relationships, perhaps for those that are weakly inbred, or for those patterns that can arise in a system of regular mating. The second line of attack is to consider the consequence of these complications for a polygenic trait and find conditions under which the relatednesses at different loci will effectively average out to some central value. Michod and Anderson suggested simple averaging over loci and this could be right. The central value could then be compared with the simple relatedness that does not distinguish between homozygotes and heterozygotes.

There is one problem with the inclusive fitness approach which may be particularly acute in the case of inbreeding. In this approach, the nature of the relationship between interactants is taken as given, whereas it may be that the postulated social actions in one generation would change the relationship between interactants in the next. For example, a gene that caused females to mate with their brothers would create a situation in the next generation where some of the brother-sister pairs were themselves the product of a brother-sister mating. The relative frequency of the 'outbred' pairs and the 'inbred' pairs matters because they will have different relatednesses. Thus, it cannot be assumed that the pattern of relatednesses between interactants is a constant independent of gene action. This is not to say that the inclusive fitness approach is useless here, only that caution is required. Maynard Smith (1980) discusses the advantages of the inclusive fitness method in tackling this and other problems.

Uyenoyama (1984) analysed exact models with specific patterns of common ancestry that involve inbreeding. She stated that the relatedness calculated from ancestry was a good guide to the models' results even when the rule strictly failed. She searched for ESSs, in which a level of altruism was non-invadable by mutations of small effect. In some cases, the non-invadable level depended on the dominance of the mutations, which implies that there is no level proof against all mutations. There is, therefore, no ESS. In other cases, a whole range of levels was strongly stable against mutations of small effect, implying the coexistence of many strongly stable ESSs. Inbreeding is a major outstanding problem for Hamilton's rule and results about the likely outcome of selection on a character would be very interesting.

Summing up the survey of derivations of Hamilton's rule, the rule has been abundantly confirmed under weak selection. Inbreeding is the only important exception reported in the literature and, while the reasons for this exception are well understood, little has been established about the likely effect on the evolution of a character.

The next, briefer survey is of definitions of relatedness. The two authoritative surveys are by Michod and Hamilton (1980), and Seger (1981), and the early work was by Hamilton (1964, 1970, 1971, 1972, 1975), Orlove (1975), and Orlove and Wood (1978). The point of these definitions was to make Hamilton's rule work, and they all involve variances and covariances of various sorts. Seger gave a particularly illuminating discussion of the assumptions under which some proposed definitions are valid. The definition in section 2 of the present paper is a slight improvement on previous definitions in three ways. Firstly, it is valid for arbitrary ploidies, thus uniting many cases in one formulation. Secondly, it allows different numbers of interactions per individual, and it allows the gene frequency of interactants to be different from the population mean. Thirdly, it allows the genotypic value to be any p-score. This draws our attention to the possibility that relatedness may be different for different alleles and shows how the relatednesses at loci in a many locus model should be combined.

Seger (1981) suggested a definition of relatedness (his R_4) which allowed actors and recipients to be genetically unrepresentative of the population. Comparison with eqn (7) shows that the reason for the complexity of his definition, and its dependence on b and c, is that it is expressed in terms of covariances around the actors' mean and recipients' mean. The simpler form of eqn (7) is achieved by taking moments about the population mean gene frequency, rather than about the actors' and recipients' means.

Michod and Hamilton (1980) gave the formula for relatedness from knowledge of common ancestry in the presence of inbreeding for diploids. They showed that all previous formulae were equivalent to it. I take this opportunity to present a particularly simple arrangement of their 'one pleomorphic coefficient'. Suppose α_{inb} and α_{outb} are the regressions of phenotype (tendency to perform the social action) on genotype (the fraction of a particular allele at a locus) within inbred actors and outbred actors, respectively, and that p_{inb} and p_{outb} are the probabilities of the actor being inbred and outbred, respectively, at that locus. Further, let φ_{inb} and φ_{outb} be the average fraction of genes at that locus in the recipient that are identical by descent with any gene at that locus in the actor, when the actor is inbred and outbred, respectively. The φs are therefore the relatednesses of inbred and outbred actors to recipients. Then, eqn (3) of Michod and Hamilton relatedness may be re-expressed as

$$\text{critical cost-benefit ratio} = \frac{2p_{inb}\alpha_{inb}\varphi_{inb} + p_{outb}\alpha_{outb}\varphi_{outb}}{2p_{inb}\alpha_{inb} + p_{outb}\alpha_{outb}}. \tag{13}$$

This is a weighted average of the relatedness of inbred actors and the relatedness of outbred actors. The idea of distinguishing these two relatednesses was suggested by Hamilton (1970). In terms of the coefficients of identity by Jacquard (1974), and as the reader can easily check from Fig. 12b,

$$\varphi_{inb} = \frac{\delta_1 + \delta_3/2}{\delta_1 + \delta_2 + \delta_3 + \delta_4}$$

and

$$\varphi_{\text{outb}} = \frac{\delta_5 + \delta_7 + \delta_8/2}{\delta_5 + \delta_6 + \delta_7 + \delta_8 + \delta_9}.$$

The weights in the weighted average are the product of three factors. The first factor is two for inbred actors and one for outbred actors, because the fate of genes in an inbred actor is twice as important from the point of view of one of his alleles. The second factor is the probability that the actor is inbred or outbred. The third factor, the regression of phenotype on genotype, is the only one that varies (explicitly at least) as a function of gene frequency and is the only one that depends on dominance. In fact it is only α_{outb} (equivalent to the α of Michod and Hamilton) that depends on gene frequency.

The form of eqn (13) makes very clear some of the properties of the critical cost-benefit ratio in the case of inbreeding. It is a constant if the two φs are equal, which means that inbred and outbred actors do not differ in their relatedness to the recipient; it does not depend on gene frequency when the heterozygotes are half-way between the homozygotes in their behaviour, for this makes α_{outb} independent of gene frequency; if the heterozygote lies between the two homozygotes in its behaviour (no over- or under-dominance) then the critical ratio lies between the two φs, because both weights are then always positive. I note that, with purely notational changes, eqn (13) is an immediate consequence of a result of Hamilton (1970).

In quantitative models, the population is divided into mutually exclusive and exhaustive groups, and the interactions have a very simple structure in which an individual affects all other group members equally. Relatedness is, in consequence, a much simpler concept and is defined as a correlation between group members. Thus, when Aoki (1982) claimed to have proved the rule 'independent of . . . inbreeding and under the action of selection', we must bear in mind that under inbreeding or selection the correlation between group members will vary between alleles and loci. Aoki had not, therefore, proved a stronger result in the quantitative case than is known for single locus models.

8. Conclusions

It may be of help to the reader to sum up the variety of material in the preceding seven sections and make some conclusions. The geometric view of relatedness was presented as a psychovisual aid in understanding the sometimes complex topics of relatedness and Hamilton's rule. Algebraic results parallel to the geometric intuition were derived in surrounding sections and these culminated in an explicit defence of Hamilton's rule as a general evolutionary principle in the study of social behaviour.

Hamilton (1964) derived his rule as a tool for understanding the evolution of social behaviour. In the central case of weak selection in an outbreeding, homogeneous population, later work has abundantly confirmed the validity of the rule as a summary of relevant population genetics models. The social behaviour of inbred and heterogeneous populations is also of great interest, however. In 1970, Hamilton extended his rule to include inbreeding, while in 1975 he explored how population structure could increase relatedness between members of the same group. The problems caused by inbreeding have been explored by Michod (1979), and Michod and Anderson (1979), and some examples worked out by Uyenoyama (1984). The fundamental problem is that the relatedness needed to predict the direction of gene frequency changes differs for dominant and recessive alleles, and depends on gene frequencies. We have seen that this same problem arises in the case of heterogeneous populations.

Neither of these difficult cases has been fully explored, and it would not be surprising if the solutions proposed by Hamilton (1972, 1975) turn out to be close to the truth most or even all of the time. On the other hand, it is also possible that there are biologically significant exceptions, cases of special interest where the solution is materially different. To the extent that Hamilton's rule is applied to species with inbreeding or with a structured population, the speedy working out of these cases is of some importance.

The rule must be judged as a success. It expresses in a precise way the ideas of altruism, spite, and selfishness. It summarizes, in a readily intelligible and accurate way, a whole host of population genetics models relevant to the likely evolution a social trait. Neither of the two serious drawbacks is understood well enough to say what other relevant factors must be included in a more embracing rule. The effects of inbreeding or of a structured population would be best summarized, if possible, in a way of defining relatedness that would make Hamilton's rule apply to those cases as well. The relative value placed on the reproduction of another, compared to the value placed on one's own, is a clear and interesting way of summarizing the action of selection on a social characteristic. This is the reason for the importance of the rule, and why we should strive to express our conclusions in terms of it.

Acknowledgements

I am very grateful to W. D. Hamilton, M. G. Bulmer, M. Ridley, and R. Dawkins for helpful comments and suggestions at various stages in the writing of this paper. This work was carried out while I held the Julian Huxley Junior Research Fellowship at Balliol College, Oxford, and a Royal Society 1983 University Research Fellowship.

References

Abugov, R. and Michod, R. E. (1981). On the relation of family structured models and inclusive fitness models for kin selection. *J. theoret. Biol.* **88**, 743–54.

Aoki, K. (1982). Additive polygenic formulation of Hamilton's model of kin selection. *Heredity,* **49**, 163–69.

Boorman, S. A. and Levitt, P. R. (1980). *The Genetics of Altruism.* Academic Press, New York.

Boyd, R. and Richerson, P. J. (1980). Effect of phenotypic variation on kin selection. *Proc. Nat. Acad. Sci. USA,* **77**, 7506–9.

Bulmer, M. G. (1985). Genetic models of endosperm evolution in higher plants. In *Evolutionary Processes and Theory* (eds S. Karlin and E. Nevo). Academic Press, New York (in press).

Cavalli-Sforza, L. L. and Bodmer, W. F. (1971). *The Genetics of Human Populations.* W. H. Freeman, San Francisco.

—— and Feldman, M. W. (1978). Darwinian selection and altruism. *Theoret. Pop. Biol.* **14**, 268–80.

Charlesworth, B. (1978). Some models of the evolution of altruistic behaviour between siblings. *J. theoret. Biol.* **72**, 297–319.

—— (1980). Models of kin selection. In *Evolution of Social Behavior: Hypotheses and Empirical Tests* (ed. H. Markl), pp. 11–26. Dahlem Konferenzen, Berlin.

—— and Charnov, E. L. (1981). Kin selection in age-structured populations. *J. theoret. Biol.* **88**, 103–19.

Charnov, E. L. (1977). An elementary treatment of the genetical theory of kin-selection. *J. theoret. Biol.* **66**, 541–50.

Cheverud, J. M. (1984). Evolution by kin selection: a quantitative genetic model illustrated by maternal performance in mice. *Evolution,* **38**, 766–77.

—— (1985). A quantitative genetic model of altruistic selection. *Behav. Ecol. Sociobiol.* **16**, 239–43.

Cohen, D. and Eshel, I. (1976). On the founder effect and the evolution of altruistic traits. *Theoret. Pop. Biol.* **10**, 276–302.

Crow, J. F. and Aoki, K. (1982). Group selection for a polygenic behavioural trait: a differential proliferation model. *Proc. Nat. Acad. Sci. USA,* **79**, 2628–31.

—— and Nagylaki, T. (1976). The rate of change of a character correlated with fitness. *Am. Nat.* **110**, 207–13.

Crozier, R. H. (1970). Coefficients of relationship and the identity of genes by descent in the Hymenoptera. *Am. Nat.* **104**, 216–17.

Dawkins, R. (1976). *The Selfish Gene.* Oxford University Press, Oxford.

—— (1979). Twelve misunderstandings of kin selection. *Zeit. Tierpsychol.* **51**, 184–200.

—— (1982). *The Extended Phenotype.* W. H. Freeman, Oxford.

Elston, R. C. and Lange, K. (1976). The genotypic distribution of relatives of homozygotes when consanguinity is present. *Ann. Human Genet.* **39**, 493–496.

Engels, W. R. (1983). Evolution of altruistic behaviour by kin selection: an alternative approach. *Proc. Nat. Acad. Sci. USA,* **80**, 515–8.

Falconer, D. S. (1981). *Introduction to Quantitative Genetics* (2nd edn). Longman, London.

Feldman, M. W. and Cavalli-Sforza, L. L. (1981). Further remarks on Darwinian selection and 'altruism'. *Theoret. Pop. Biol.* **19**, 251–60.

Fisher, R. A. (1958). *The Genetical Theory of Natural Selection.* Dover, New York.

Grafen, A. (1979). The hawk-dove game played between relatives. *Anim. Behav.* **27**, 905–7.

—— (1984). Natural selection, kin selection and group selection. In *Behavioural Ecology, 2nd edn.* (eds J. R. Krebs and N. B. Davies). pp. 62–84. Blackwell Scientific Publications, Oxford.

Haldane, J. B. S. (1955). Population genetics. *New Biol.* **18**, 34–51.

Hamilton, W. D. (1963). The evolution of altruistic behaviour. *Am. Nat.* **97**, 354–6.

—— (1964). The genetical evolution of social behaviour. I and II. *J. theoret. Biol.* **7**, 1–52.

—— (1967). Extraordinary sex ratios. *Science,* **156**, 477–88.

—— (1970). Selfish and spiteful behaviour in an evolutionary model. *Nature,* **228**, 1218–20.

—— (1971). Selection of selfish and spiteful behaviour in some extreme models. In *Man and Beast: Comparative Social Behaviour* (eds J. F. Eisenberg and W. S. Dillon), pp. 55–91. Smithsonian Institution Press, Washington.

—— (1972). Altruism and related phenomena, mainly in social insects. *Ann. Rev. Ecol. System.* **3**, 193–232.

—— (1975). Innate social aptitudes of man: an approach from evolutionary biology. In *Biosocial Anthropology* (ed. R. Fox), pp. 133–55. John Wiley and Sons, New York.

Harpending, H. C. (1979). The population genetics of interactions. *Am. Nat.* **113**, 622–30.

Jacquard, A. (1974). *The Genetic Structure of Populations.* Springer, New York. (Translated from the French by D. and B. Charlesworth. Volume 5 in the Biomathematics Series.)

Karlin, S. and Matessi, C. (1983). Kin selection and altruism. *Proc. Roy. Soc. Lond. Ser. B,* **219**, 327–53.

Knowlton, N. and Parker, G. A. (1979). An evolutionarily stable strategy approach to indiscriminate spite. *Nature,* **279**, 419–21.

Levitt, P. R. (1975). General kin selection models for genetic evolution of sib altruism in diploid and haplodiploid species. *Proc. Nat. Acad. Sci. USA,* **72**, 4531–5.

Lewontin, R. C. (1974). *The Genetic Basis of Evolutionary Change.* Columbia University Press, New York.

Malécot, G. (1969). *The Mathematics of Heredity.* W. H. Freeman, San Francisco. (Translated from the French by D. M. Yermanos.)

Matessi, C. and Jayakar, S. D. (1976). Conditions for the evolution of altruism under Darwinian selection. *Theoret. Pop. Biol.* **9**, 360–87.

Maynard Smith, J. (1980). Models of the evolution of altruism. *Theoret. Pop. Biol.* **18**, 151–9.

—— (1982). The evolution of social behaviour – a classification of models. In *Current Problems in Sociobiology* (eds Kings College Sociobiology Group). pp. 29–44. Cambridge University Press, Cambridge.

Michod, R. E. (1979). Genetical aspects of kin selection: effects of inbreeding. *J. theoret. Biol.* **81**, 223–33.

—— (1982). The theory of kin selection. *Ann. Rev. Ecol. System.* **13**, 23–55.

—— and Anderson, W. W. (1979). Measures of genetic relationship and the concept of inclusive fitness. *Am. Nat.* **114**, 637–47.

—— and Hamilton, W. D. (1980). Coefficients of relatedness in sociobiology. *Nature,* **288**, 694–7.

Nevo, E. (1978). Genetic variation in natural populations: patterns and theory. *Theoret. Pop. Biol.* **13**, 121–77.

Noonan, K. M. (1981). Individual strategies of inclusive-fitness-maximizing in *Polistes fuscatus* foundresses. In *Natural Selection and Social Behaviour: Recent Research and New Theory* (eds R. D. Alexander and D. W. Tinkle), pp. 18–44. Chiron Press, New York.

O'Donald, P. (1982). The concept of fitness in population genetics and sociobiology. In *Current Problems in Sociobiology* (eds Kings College Sociobiology Group), pp. 65–85. Cambridge University Press, Cambridge.

Orlove, M. J. (1975). A model of kin selection not invoking coefficients of relationship. *J. theoret. Biol.* **49**, 289–310.

—— and Wood, C. L. (1978). Coefficients of relationship and coefficients of relatedness in kin selection: a covariance form for the rho formula. *J. theoret. Biol.* **73**, 679–86.

Pamilo, P. and Crozier, R. H. (1982). Measuring genetic relatedness in natural populations: methodology. *Theoret. Pop. Biol.* **21**, 171–93.

Price, G. R. (1970). Selection and covariance. *Nature,* **227**, 520–1.

—— (1972). Extension of covariance selection mathematics. *Ann. Human Genet.* **35**, 485–90.

Queller, D. C. (1984a). Models of kin selection on seed provisioning. *Heredity,* **53**, 151–65.

—— (1984b). Kin selection and frequency dependence: a game theoretic approach. *Biol. J. Linn. Soc.* **23**, 133–43.

Robertson, A. (1966). A mathematical model of the culling process in dairy cattle. *Anim. Prod.* **8**, 95–108.

—— (1968). The spectrum of genetic variation. In *Population Biology and Evolution* (ed. R. C. Lewontin), pp. 5–16. Syracuse University Press, New York.

Rushton, J. P., Russell, R. J. H., and Wells, P. A. (1984). Genetic similarity theory: beyond kin selection. *Behav. Genet.* **14**, 179–93.

Scudo, F. M. and Ghiselin, M. T. (1975). Familial selection and the evolution of social behaviour. *J. Genet.* **62**, 1–31.

Seger, J. (1976). Evolution of responses to relative homozygosity. *Nature,* **262**, 578–80.

—— (1981). Kinship and covariance. *J. theoret. Biol.* **91**, 191–213.

Uyenoyama, M. K. (1984). Inbreeding and the evolution of altruism under kin selection: effects on relatedness and group structure. *Evolution,* **38**, 778–95.

—— and Feldman, M. W. (1981). On relatedness and adaptive topography in kin selection. *Theoret. Pop. Biol.* **19**, 87–123.

—— and —— (1982). Population genetic theory of kin selection. II The multiplicative model. *Am. Nat.* **120**, 614–27.

——, ——, and Mueller, L. D. (1981). Population genetic theory of kin selection: multiple alleles at one locus. *Proc. Nat. Acad. Sci. USA,* **78**, 5036–40.

Wade, M. J. (1982). The effect of multiple inseminations on the evolution of social behaviours in diploid and haplodiploid organisms. *J. theoret. Biol.* **95**, 351–68.

—— (1985). Soft selection, hard selection, kin selection and group selection. *Am. Nat.* **125**, 61–73.

Wilson, D. S. (1975). A theory of group selection. *Proc. Nat. Acad. Sci. USA,* **72**, 143–6.

Wright, S. (1968, 1969, 1977, 1978). *Evolution and the Genetics of Populations,* Volumes 1–4. University of Chicago Press, Chicago.

Yokoyama, S. and Felsenstein, J. (1978). A model of kin selection for an altruistic trait considered as a quantitative character. *Proc. Nat. Acad. Sci. USA,* **75**, 420–2.

Measures of sexual selection

WILLIAM J. SUTHERLAND

Sexual selection often appears to act more strongly upon one sex than it does on the other. Sexual selection also appears more important in some species than in others. There have been many attempts to explain and describe these discrepancies. It has been suggested that the variation in the importance of sexual selection can be explained by studying either the variation in relative parental investment (Trivers 1972) or the variation in the operational sex ratio (Emlen 1976). Additional measures attempt to calculate the consequences of sexual selection for the breeding success of different individuals. These include the ratio of maximal breeding success (Ralls 1977), variance in mating success (Wade and Arnold 1980) and relative variance in reproductive success (Payne 1979). All of these measures have been developed independently of each other. In this review I would like to consider how these measures are related.

I shall also suggest a new measure of the expected strength of sexual selection. This measure is developed from ideas of foraging behaviour. In the same way as the food intake can be restricted by the handling time (Holling 1959), the mating success can be restricted by the time necessary for mating and caring for the offspring. With this approach, the essential difference between males and females lies in their different mating times. This model leads to the conclusion that the strength of sexual selection can be assessed by measuring the proportion of time spent seeking mates.

The current approach to sexual selection is to measure variance in mating success (e.g., Wade 1979; Wade and Arnold 1980) and I will question the usefulness of this approach. The advantage of the new model proposed in this paper is that it has a clear algebraic justification and it can show the relationship between other measures currently in use.

A model of sexual selection

The strength of sexual selection is the degree to which different genotypes have different fitnesses due to differences in the frequency or quality of mates; I shall develop a model to see what factors influence this. Sexual selection may often act through differences in mate quality (e.g., O'Donald 1983), but for simplification I will ignore this aspect and just consider the extent to which different genotypes differ in mating rate. In common with many other studies and most field work, I will assume mating rate is a measure of fertilization rate and will ignore differences in mating rate due to sperm competition. This model, therefore, is not a complete description of all aspects of sexual selection.

Parker (1979) introduced the concept of the aptitude for encounter, which is the number of matings acquired in a given time spent seeking mates. A high value of the aptitude for encounter may be either because the individual is efficient at seeking out mates or because it is good at persuading the other sex to mate with it. For this model, it makes no difference whether mate choice is 'active' or 'passive' (Parker 1982). If different genotypes vary in their ability to acquire matings it will be because they vary in their aptitude for encounter and the source of this difference may vary between species. This difference in aptitude for encounter may be because individuals differ in plumage (Andersson 1982), dominance status (Le Boeuf 1974), size (reviewed in Ridley 1983), courtship ability (Maynard Smith 1956), courtship feeding performance (Thornhill 1976), fighting ability (Clutton-Brock *et al.* 1982), or even, in a satin bowerbird *Ptilinorhynchus violaceus*, on how good an individual is at collecting blue items for the bower (Borgia 1985). Thus, I assume that the strength of sexual selection is the relationship between the mating rate and the genetic basis for the aptitude for encounter.

What are the consequences of a change in the aptitude for encounter in terms of the number of matings? Assume that individuals are able, over their lifespan, to devote a certain amount of time, T, to activities related to obtaining matings; part of this time will be spent seeking mates and the rest will be associated more directly with mating. Each mating leaves the animal unable to seek other partners for a time, H. For these purposes, mating covers the time necessary for activities such as mate guarding, copulation, parental care, and replacing gametes spent during the mating. Assume that mates are encountered at random at a rate λ during the time spent seeking mates. If a mutant is n times more capable at acquiring mates than is an average individual, it will find mates at a rate $n\lambda$. We can calculate the number of matings F' which we expect it to obtain in time T (after Holling 1959)

$$\frac{F'}{T} = \frac{n\,\lambda}{1 + n\,\lambda\,H} \tag{1}$$

The rate at which the average individual (for whom $n = 1$) encounters potential mates whilst searching can be found from the rearrangement of eqn (1)

$$\lambda = \frac{1}{(T/F')-H} \tag{2}$$

where F is the number of matings obtained by an average individual. Substituting eqn (2) into eqn (1) gives

$$\frac{F'}{T} = \frac{n\,/\,\{(T\,/\,F')-H\}}{1 + n\,H\,/\,\{(T/F)-H\}} \tag{3}$$

or

$$\frac{F'}{F} = \frac{1}{FH\,/\,T + (T - FH)\,/\,nT} \tag{4}$$

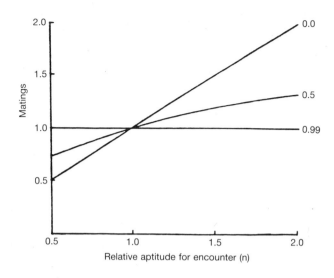

Fig. 1. The extent to which different aptitudes for encounter result in different numbers of matings. The relationship is shown for different fractions of time spent seeking $(T - FH)/T$. The steeper the slope, the stronger the sexual selection.

where F'/F is the relative mating success of the mutant. Figure 1 shows the relationship between this and n, the relative aptitude for encounter for the mutant for various fractions of time spent seeking $(T - FH)/T$.

As stated before, the strength of sexual selection is the relationship between the genetic basis for the aptitude for encounter and the rate at which fertilizations (or matings) are acquired. Figure 1 shows how the rate at which matings are acquired depends upon the relative aptitude for encounter. The relative aptitude for encounter expresses plumage colour, calling ability, fighting ability, blue-item-collecting ability, or whatever, so that Fig. 1 is the relationship between these qualities and mating success. If this relationship is steep then sexual selection is strong: the best fighters or the most attractive get far more mates. If the slope is shallow then sexual selection is weak. The steepness of the slope is determined by differentiating eqn (4).

$$\frac{d\,(F'/F')}{dn} = \frac{(T - FH)T}{(T + nFH - FH)^2} \tag{5}$$

As mutations affecting mate-finding ability are likely to be small, most mutants will be only slightly better or only slightly worse at finding mates than average, i.e., $n \simeq 1$. Under this condition eqn (5) simplifies to

$$\frac{d\,(F'/F)}{dn} \simeq \frac{T - FH}{T} \tag{6}$$

As F is the number of matings obtained by an average member of the population and H is the time available for reproduction which is taken up by one mating, it is clear that $(T - FH)/T$ is the fraction of time an average individual spends in seeking mates.

If a particular mutant is successful and spreads then the population will acquire the encounter rate of the mutant, i.e., n will become equal to one. If there is frequency-dependent selection acting upon the character then n will exceed one when it is rare, but may even be less than one when it is abundant.

It thus seems that the fraction of time spent searching for mates, i.e., $(T - FH) / T$ provides a measure of the strength of sexual selection. When most of an individual's time is devoted to seeking mates then genotypes differing in the aptitude for encounter will have very different numbers of mates: sexual selection will be strong (see Fig. 1). When a low proportion of the time is spent seeking then genotypes may differ in their aptitude for encounter, but still have similar numbers of matings; i.e., sexual selection will be weak. In many species it is the female which spends most time producing and caring for the young and so they spend less time seeking. As a result sexual selection will act less strongly upon them than it will upon males which is probably why males are often brightly coloured, or display, or sing. In other species, such as phalaropes *Phalaropus* sp., pipefish and seahorses (Syngathidae), and the frog, *Dendrobates aurata*, the mating time for males is longer and sexual selection may be expected to act more strongly upon the females – again probably accounting for their brighter colours (Trivers 1972).

The fraction of time spent seeking might be assessed by counting the number (N_s) of individuals seeking mates at any one time and the number (N_f) involved in activities associated with matings (i.e., mate guarding, mating, parental care, and replacing gametes); the required fraction is then $N_s /(N_s + N_f)$. This will still not be easy to measure. The fraction of time spent seeking may vary over the year or between seasons or habitats. The strength of sexual selection will vary in unison.

This fraction of time can be expressed more usefully as $F / T \lambda$, that is, the strength of sexual selection is dependent upon the mean mating rate, F / T, divided by the mean rate at which mates are encountered by a seeking individual, λ.

Comparisons with previous approaches

The model just outlined suggests that the fraction of time spent searching may be used as a measure of the strength of sexual selection. I will now compare this with previous measures (summarized in Table 1).

It is necessary to distinguish between indices of the strength of sexual selection acting upon one sex (Wade and Arnold 1980; this paper, eqn 5) and those indices which describe the ratio of the strength of sexual selection acting upon the two sexes (Trivers 1972; Emlen 1976; Ralls 1977; Payne 1979; this paper, eqn 7). These indices should be the same if there is no sexual selection acting upon one of the sexes.

RELATIVE TIME SPENT SEEKING MATES

The ratio of the average amount of time spent seeking by males to the average amount of time spent seeking by females is

$$\frac{T - F_m H_m}{T - F_f H_f}. \tag{7}$$

According to the model outlined earlier this will be equivalent to the ratio of the strengths of sexual selection.

THE OPERATIONAL SEX RATIO

Emlen (1976) suggested measuring the operational sex ratio, which is the ratio of seeking males to seeking females. For a given absolute sex ratio S,

$$\text{Operational sex ratio} = \frac{T - F_m H_m}{T - F_f H_f} \; S. \tag{8}$$

This is eqn (7) multiplied by the absolute sex ratio S. Thus, if the absolute sex ratio is unity then the operational sex ratio is an approximate measure of the ratio of the strengths of sexual selection between the sexes and, contrary to Emlen (1976), is not a straightforward measure of sexual selection. Those sex-linked characters which enable one sex to compete against themselves such as the tusks of the narwhal, *Monodon monoceros*, or the lekking behaviour of ruffs, *Philomachus pugnax*, may be expected in species such as these in which there is a highly biased operational sex ratio. The operational sex ratio does seem very useful and comparatively easy to measure, yet seems to be little used.

RELATIVE PARENTAL INVESTMENT

Trivers' (1972) analysis is based upon his concept of parental investment which he defines as 'any investment by the parent in an individual offspring that increases the offspring's chance of surviving (and hence reproductive success) at the cost of the parent ability to invest in other offspring'. When a member of one sex invests a time, H in one mating it is losing the opportunity of finding $H \lambda$ other matings. $H \lambda$ can thus be considered a crude measure of parental investment. Substituting the term for λ given in eqn (2) produces

$$\text{Parental investment (P.I.)} = \frac{H}{(T/F') - H}. \tag{9}$$

Trivers (1972) suggested that 'what governs the operation of sexual selection is the relative parental investment of the sexes in their offspring'. By using eqn (9), this can be formulated as

$$\frac{\text{Female P.I.}}{\text{Male P.I.}} = \frac{H_f \{ (T/F_m) - H_m \}}{H_m \{ (T/F_f) - H_f \}} \tag{10}$$

Table 1
Summary of the suggested relationships between some models of sexual selection. Note that the first expression of each is the same.

Relationship	Model		
Relative time spent seeking mates	$\dfrac{T - F_m H_m}{T - F_f H_f}$		
Operational sex ratio	$\dfrac{T - F_m H_m}{T - F_f H_f}$	$\cdot$	S
Relative parental investment	$\dfrac{T - F_m H_m}{T - F_f H_f}$	$\cdot$	$S \; \dfrac{H_f}{H_m}$
Relative variance in reproductive success	$\dfrac{T - F_m H_m}{T - F_f H_f}$	$\cdot$	$\dfrac{1}{S} \left(\dfrac{T - F_m H_m}{T - F_f H_f} \right)$
Relative standard deviation in reproductive success	$\dfrac{T - F_m H_m}{T - F_f H_f}$	$\cdot$	$\sqrt{\dfrac{1}{S}}$
Ratio of maximal breeding	$\dfrac{T - F_m H_m}{T - F_f H_f}$	$\cdot$	$\dfrac{(T - F_m H_m)\, H_m}{(T - F_f H_f)\, H_f}$

T = Total time; F = mean number of matings; H = mating time; S = sex ratio. See text for the strict definitions of these terms.

or

$$\frac{\text{Female P.I.}}{\text{Male P.I.}} = \left(\frac{T - F_m H_m}{T - F_f H_f} \right) \cdot \left(\frac{F_f H_f}{F_m H_m} \right). \tag{11}$$

This expression will give something like the same results as the operational sex ratio [eqn (8)]; indeed, the first term is identical and F_f / F_m is equivalent to the sex ratio, S. Thus, the actual results given by this formulation of relative parental investment will deviate from those of the operational sex ratio by an amount H_f / H_m.

RELATIVE VARIANCE IN REPRODUCTIVE SUCCESS

Payne (1979) mentioned using variance in male reproductive success divided by variance in female reproductive success as a measure of sexual selection. For this section, I assume that all variance is caused by chance. Modifying eqn (3.3.7) of Cox (1962) gives, as a good approximation, the relationship between variance in reproductive success and mating time

$$\text{Variance} \quad = \quad \frac{\lambda\, T}{(1 + \lambda\, H)^3} \tag{12}$$

Thus, the relative variance becomes

$$\frac{\lambda_m\, /\, (1 + \lambda_m H_m)^3}{\lambda_f\, /\, (1 + \lambda_f H_f)^3} \tag{13}$$

which, with the help of eqn (2), gives us

$$\frac{F_m}{F_f} \cdot \left(\frac{T - F_m H_m}{T - F_f H_f} \right)^2. \tag{14}$$

The term in brackets is the relative strength of sexual selection of eqn (7). The ratio of matings ($F_m\, /\, F_f$) is the reciprocal of the actual sex ratio. Thus, it seems that, if the actual sex ratio is unity, then relative variance in reproductive success estimates the sex difference of the squares of the strengths of sexual selection as measured by my temporal model.

RELATIVE STANDARD DEVIATION IN REPRODUCTIVE SUCCESS

As pointed out to me by Alan Grafen, eqn (14) can be made to relate even more closely to the others by measuring the standard deviations, which is the square root of the variance. This gives

$$\sqrt{\frac{1}{S}} \cdot \frac{T - F_m H_m}{T - F_f H_f} . \tag{15}$$

If the sex ratio S (or $F_f\, /\, F_m$) is unity, then this also measures the ratio of strengths of sexual selection.

RATIO OF MAXIMAL BREEDING

Ralls (1977) suggested 'the intensity of intrasexual selection in a species should be proportional to the ratio of the lifetime number of offspring reared by a highly successful male compared to the number born by a highly successful female in her lifetime'. The maximum number of offspring that can be produced is $T\, /\, H$ which I will consider to be the number produced by the most successful individuals – this is an overestimate which is likely to deviate most from reality when fertilization time is very low. Hence, Ralls' measure approximately equals

$$\frac{1\, /\, H_m}{1\, /\, H_f} \quad \text{or} \quad \frac{H_f}{H_m} \tag{16}$$

This deviates markedly from the other measures.

VARIANCE WITHIN ONE SEX

Wade and Arnold (1980) have argued that variance in lifetime reproductive success is a direct measure of the intensity of sexual selection. There is a lucid review of this subject by Clutton-Brock (1983) – the verbal presentation of which inspired the work presented here. The aim of this section is to question the usefulness of measuring variance in reproductive success in the study of sexual selection. I shall suggest that it is neither a measure of sexual selection, nor is it the most direct evidence for the potential for sexual selection, indeed, that it is not necessarily evidence for the operation of sexual selection in a particular case at all.

Following the pioneering work of Bateman (1948), various workers have shown that males have more variance in mating success than do females (Clutton-Brock *et al.* 1982; Fincke 1982; Howard 1978) and that this effect is strongest in those species in which the male provides minimal care of the offspring (Payne and Payne 1979). From this it has been argued that, in general, males must compete with each other for the opportunity to mate with females and thus experience stronger sexual selection (Trivers 1972).

My major objection to using variance in mating success is that chance may play an important role and that the contribution of chance may vary consistently between species and sexes (Sutherland 1985). A group which invests little time in each mating is likely, purely by chance, to show great variance; a group which invests considerable time in each mating will show little variance because an individual would have to be exceptionally lucky to find many mates. For example, if each mating takes up 20 per cent of the time available then it would be impossible to find six mates and five mates would each have to be found instantaneously.

This can be shown more formally by incorporating mating time into a Poisson distribution as illustrated in Fig. 2 (Sutherland 1985). If mating time is negligible and matings occur at random then the frequency of matings will conform to a simple Poisson distribution (Fig. 2a). If mating time takes up quite a long time then there will be much less variance in mating success (Fig. 2b,c).

In many species the mating time will be longer for females as they produce the larger gametes and often do a disproportionate share of looking after the young. We may expect females to show little variance and males to have a large variance simply because of the difference in mating times. Studies of this sort also have to exclude other possible sources of variance. With a small effective population size, the sex ratio is likely to deviate from unity by chance within each subpopulation and so produce variance in the number of matings. Mortality may also be an important source of variance (Sutherland and Banks, in press).

Bateman (1948) led people astray by comparing male and female variance rather than considering them separately. A more direct test is to see if the mating success of males in Bateman's experiment deviates from a binomial distribution; it does not (Sutherland 1985). So even though Bateman's study is the standard text-book example of non-random mating

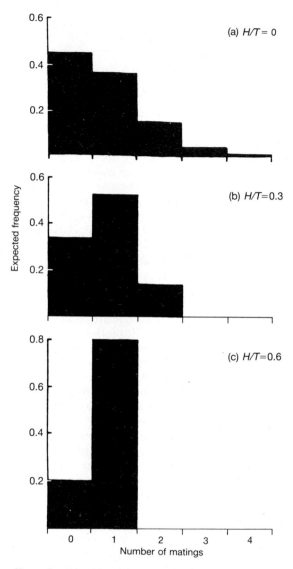

Fig. 2. The effect of mating time upon the amount of variation in mating success determined by chance. Each mating takes (a) negligible time, (b) 30 per cent of the total time available, (c) 60 per cent of total time (after Sutherland, 1985).

there is no evidence for deviation from randomness. There is no real evidence that certain individuals were better competitors or favoured in some way.

Wade and Arnold (1980) suggested that the intensity of sexual selection upon males can be calculated as the variance in male mating rate. Their argument is a modification of the result derived by Crow (1958). The rate at which mean fitness can change depends upon the strength of selection and the available diversity of genotypes upon which this selection can act – this is Fisher's (1930) fundamental theorem of natural selection. If fitness is completely heritable, then the variation in fitness is an index of the rate at which mean fitness will change. It does not matter if the variation in fitness is caused by a wide diversity of genotypes combined with weak selection or by a limited range of genotypes and strong selection. Thus, it predicts a rate of change for the population (Crow referred to this as 'total selection intensity') and not the selection acting upon a given genotype (which Wade and Arnold call the response to selection). For a known range of genotypes, the variance in number of matings is a measure of the sexual selection (subject to an assumption described below), but without knowledge of the range of genoptypes it means little to use variance in the number of matings as an index of the strength of sexual selection.

Furthermore, as Crow (1958) pointed out, this assumes fitness 'is completely heritable, that is, each offspring has exactly the average of his parents' fitness'. Phenotypic influences are obviously very important; genetically similar individuals may, through chance environmental influences, have different phenotypes and hence different numbers of matings. Attempts to measure heritability have shown that characters which have little effect on fitness may have high heritabilities, but those characters which have important consequences for fitness (e.g., any related to sexual selection) have low heritabilities (Falconer 1964). This is simply because it is not possible, over the long term, to have both high heritability and strong selection upon phenotypes as the population would soon lose all its additive genetic variance. The Wade and Arnold approach would be justified if the same fraction of total variance is always heritable. However, it seems that their measure includes some unavoidably random variation and that this random variation is different between sexes and between species, and is by definition uninheritable.

I do not deny that variance in reproductive success is in some way related to the strength of sexual selection. If some individuals have the opportunity to mate a disproportionate number of times, then there is likely to be both variance in mating success and competition to be one of the successful individuals. However, it seems that the fraction of time spent seeking for mates is a more direct estimate of the opportunity to mate a disproportionate number of times. Of all the available measures, the operational sex ratio is probably the easiest to calculate and interpret.

Acknowledgements

Mike Banks, Richard Dawkins, Alan Grafen, Mark Ridley, Dave Thompson, and Paul Ward, responded to various ideas with indispensible encouragement, criticism, and laughter. Clive Anderson gave an insight into renewal theory. Duncan Brooks rewrote most of the sentences and equations. My greatest debt is to Geoff Parker who made a considerable contribution to this paper, but with typical modesty refused joint authorship.

References

Andersson, M. (1982). Female choice selects for extreme tail length in a widowbird. *Nature, 299*, 818–20.

Bateman, A. J. (1948) Intrasexual selection in *Drosophila Heredity, 2*, 349–68.

Borgia, G. (1985). Bower quality, number of decorations and mating success of male satin bowerbirds (*Ptilonorhychus violaceus*): an experimental analysis. *Anim. Behav. 33*, 266–71.

Clutton-Brock, T. H. (1983). Selection in relation to sex. In *Evolution from Molecules to Men* (ed. D. S. Bendall), pp. 457–86. Cambridge University Press, Cambridge.

——, Guiness, F. E., and Albon, S. D. (1982). *Red Deer. Behaviour and Ecology of Two Sexes*. Edinburgh University Press, Edinburgh.

Cox, D. R. (1962). *Renewal Theory*. Methuen, London.

Crow, J. F. (1958). Some possibilities for measuring selection intensities in man. *Human Biol. 30*, 1–13.

Emlen, S. (1976). Lek organisation and mating strategies in the bullfrog. *Behav. Ecol. Sociobiol. 1*, 283–313.

Falconer, D. S. (1964). *Introduction to Quantitative Genetics*. Longman, London.

Fisher, R. A. (1930). *The Genetical Theory of Natural Selection*. Clarendon Press, Oxford.

Fincke, O. M. (1982). Lifetime mating success in a natural population of the damselfly, *Enallagma hageni* (Walsh) (Odonata: Coenagrionidae) *Behav. Ecol. Sociobiol. 10*, 293–302.

Holling, C. S. (1959). Some characteristics of simple types of predation and parasitism. *Can. Ent. 91*, 385–98.

Howard, R. D. (1978). The evolution of mating strategies in bullfrogs *Rana cetesbiana. Evolution, 32*, 850–71.

Le Boeuf, B. J. (1974). Male-male competition and reproductive success in elephant seals. *Am. Zool. 14*, 163–76.

Maynard Smith, J. (1956). Fertility, mating behaviour and sexual selection in *Drosophila subobscura. J. Genet. 54*, 261–79.

O'Donald, P. (1983). *The Arctic Skua*. Cambridge University Press, Cambridge.

Parker, G. A. (1979). Sexual selection and sexual conflict. In *Sexual Selection and Reproductive Competition in Insects* (eds M. S. Blum and N. A. Blum), pp. 123–66. Academic Press, London.

—— (1982). Phenotype limited evolutionary stable strategies. In *Current Problems in Sociobiology* (ed. Kings College Sociobiology Group), pp. 173–201. Cambridge University Press, Cambridge.

Payne, R. B. and Payne, K. (1979). Social organisation and mating success in local song populations of village indigobirds *Vidua chalybeata. Zeit. Tierpsychol. 45*, 113–73.

Payne, R. B. (1979). Sexual selection and intersexual differences in variance of breeding success. *Am. Nat. 114*, 447–52.

Ralls, K. (1977). Sexual dimorphism in mammals: avian models and unanswered questions. *Am. Nat. 111*, 917–38.

Ridley, M. (1983). *The Explanation of Organic Diversity*. Oxford University Press, Oxford.

Sutherland, W. J. (1985). Chance can produce a sex difference in variance in mating success and account for Bateman's data. *Anim. Behav. 34*, 1349–1352.

—— and Banks, M. (in press). Chance as a source of variance in male reproductive success. In *Reproductive Success* (ed. T. H. Clutton-Brock). University of Chicago Press, Chicago.

Thornhill, R. (1976). Sexual selection and nuptual feeding in *Bittacus apicalis* (Insecta: Macroptera). *Am. Nat.* **110**, 529–48.

Trivers, R. L. (1972). Parental investment and sexual selection. In *Sexual Selection and the Descent of Man* (ed. B. Campbell) pp. 139–79. Aldine, Chicago.

Wade, M. J. (1979). Sexual selection and variance in reproductive success. *Am. Nat.* **114**, 742–7.

—— & Arnold, S. J. (1980). The intensity of sexual selection in relation to male sexual behaviour, female choice, and sperm precedence. *Anim. Behav.* **28**, 446–61.

Issues in brain evolution

HARRY J. JERISON

The fossil record of the vertebrate brain goes back about 400 million years, and is represented by several thousand endocranial casts or endocasts (Edinger 1975). This important evidence is clearest for the avian and mammalian brain, because in most species in these groups the internal surface of the cranium mirrors in remarkable detail the external surfaces of the brain. In mammals, the accuracy of an endocast as a model of a brain is excellent for most species; the exceptions are mainly the large-brained hominoid primates, the cetaceans, and the elephants. Brain size is reasonably well estimated by endocranial volume in all mammals, but the details of the convolutional pattern are usually lost in those large-brained species.

The evidence from fossils is about the external configuration of the brain. What does such information about the exterior tell us about the internal structure and functions? There are two kinds of answers, both based on research on living brains. Firstly, there is the correlation of superficial structures with specialized behavioural and neurophysiological functions, which has led to our knowledge of the localization of functions in the brain. Secondly, there are biometric relationships between brain size and statistical aspects of the brain's work. Although sometimes misunderstood as supporting a simple-minded notion of 'mass action' by the brain, this second approach is, in fact, consistent with a developing consensus about the integrative action of localized systems and the multiple representation of functions.

The first approach leads to functional mappings, identification of functional areas, and suggestions about specialization of brains in terms of functional areas. If convolutional patterns present in fossil endocasts are similar to those in living species, the simplest ('uniformitarian') assumption is that functional localization was similar in the fossil and living species. The main evolutionary contribution of this approach has been to confirm the expectation that present specializations of the brain in various species are probably similar to past specializations in the brain of ancestors of those species, with some evidence for anagenetic evolution of specialization.

In most mammalian endocasts, whether from living or fossil species, the convolutional patterns of the cerebrum are identifiable. They are similar enough for ancestral and descendant species so that particular gyri and sulci can be given the same names, such as 'lateral', 'arcuate', 'ectosylvian' sulci, the 'rhinal' fissure, and so forth (Radinsky 1979). Every structure visible on these petrified endocasts can be identified and named as if it were a structure on a brain, although there have been occasional debates on whether a particular gyrus or sulcus has been correctly identified in ancestral and descendant species (see Falk 1980; Holloway 1981b). Major parts of the brain, such as the cerebral hemispheres and the cerebellum, are easily identified in most mammalian endocasts viewed as fossil brains. There are often questions raised about the small size of some fossil endocasts, but they must always be understood in relation to body size.

This leads to the second line of evidence available from external structure, evidence on the evolution of brain size. About a decade ago, I published a monograph on brain evolution (Jerison 1973) in which the primary data were the sizes of fossil endocasts. These data, appropriately interpreted, provided unique evidence on the evolution of intelligence. There is now more evidence and there are new perspectives. Much of this review is devoted to a reconsideration of issues raised in the monograph as well as issues raised by its critics.

The most interesting, most difficult, and most controversial issues arise from the problems of defining and understanding intelligence as a trait that evolved. Brain size and brain/body relationships, which can be estimated in fossil as well as living animal species, are especially important for understanding the relationship between brain and intelligence, and have been reviewed in several new analyses. Evolutionary theory, applied in new ways to the analysis of individual and social behaviour, and to the rate and direction of evolution, may be applicable in related ways to the brain's evolution. Finally, new evidence on human evolution and on neurobiology has modified at least some of the conclusions about the evolution of the brain.

Biological intelligence

Major texts and monographs on intelligence in biological and evolutionary perspective have been published during the past decade (Bindra 1976; Eccles 1979; Griffin 1976; Lumsden and Wilson 1981; Macphail 1982; Oakley and Plotkin 1979; Passingham 1982; Walker 1983) and there have been many shorter reviews (Bitterman 1975; Herman 1980; Jerison 1982b; Parker and Gibson 1979; Plotkin 1983; Warren 1977). The literature is extensive and impressive, although its limitations are evident in the frequency with which the problem of definition is raised. There is an incomplete consensus on the nature of intelligence, even of human intelligence, and defining animal intelligence remains a fundamental problem for the biology of mind.

From Aristotle to Darwin to our own time, naturalists have identified intelligence with the capacity to learn and to adapt to new situations. The programme of comparative psychology is in this classic tradition, comparing species in their performance in mazes, problem boxes, and other standardized tasks to measure their learning capacities (Bitterman 1975; but see also Macphail 1982; Warren 1977). Others, using Piaget's (1971) ideas that ontogenetic stages in the growth of the child's mind recapitulate a phylogeny of mind, have proposed a Piagetian framework to identify cognitive grades in the behaviours of living primates (Parker and Gibson 1979; but see also Jerison 1982c, and other commentaries). Problem solving ability is often suggested as a potential measure (Passingham 1982), and intelligence is sometimes equated with inquisitiveness or curiosity (Glickman and Sroges 1966; Parker 1974).

Human intelligent behaviour has clearly been the model for our ideas about intelligence as a biological trait in other species. That a behaviour pattern requires intelligence when performed by humans is no guarantee that the behaviour pattern in other species also implies intelligence, however, nor that its absence indicates low intelligence. Macphail (1982) has reviewed an extensive literature related to this issue. His standards for rejection of a null hypothesis of 'no difference' among species are perhaps stringent, but it is significant that, in almost every category of objective behavioural evidence that he examined, he could not conclude that a difference in intelligence among major groups of species had been demonstrated unequivocally. Only in the presence or absence of human language did he see strong evidence that one species (*Homo sapiens*) was differentiable from other animal species on a scale of mental capacity.

Part of the problem may be inherent in the human origin of the idea of intelligence. Intelligence is a quantitative idea about behavioural capacity and not about specialized brain/behaviour systems that would have evolved as specialized adaptations in the natural selection of species for their niches. In the concept of intelligence there is no recognition that behaviour that is complex in the human species may be a simple, 'unintelligent' adaptation to an unusual niche in another species. The considerable abilities of laboratory rats at maze learning, for example, may be part of their specialized adaptations for navigating the elaborate systems of burrows that are their natural habitat (Barnett 1963; Calhoun 1963; Warren 1957). Rats were 'prepared' (Seligman 1970) by their evolutionary histories to learn mazes, according to this view, and their good performance in mazes does not necessarily imply great learning ability in all tasks.

Behaviours analysed to measure intelligence in different species must be comparable if they are to measure the species. Can there be truly comparable behaviours? Should we not anticipate some interaction of any behaviour system with species-typical adaptations? To the extent that brain/behaviour systems evolved under natural selection, they should be understood in the context of an adaptation to an environmental niche. There must be a context within which every specialized behaviour is organized, including behaviour used to measure intelligence. If a species is adapted for certain behavioural solutions to an environmental problem, it will measure high on an 'animal IQ' test based on those behaviours. The example of maze learning suggested that rats may be more 'stupid' than their excellent performance in mazes suggests, because they are specialized to do well in mazes.

To illustrate how a reverse bias about intelligence can be introduced by the interaction of species-typical adaptations with a behavioural test, consider exploratory behaviour. This kind of behaviour, 'curiosity', is a hallmark of intelligence in children in Western classrooms, though it may signify willfulness and lack of discipline in other parts of the world. It may be an unnatural trait in most animal species, because exploring a natural environment can be fatal; timidity in facing novelty is a preferred route to survival. There are, furthermore, systematic differences among species depending on their adaptive modes. Predators are likely to be timid in ways quite different from prey and members of social species are likely to be systematically different from non-social ones in their 'curiosity' about conspecifics. Gibbons and spider monkeys seemed to be peculiarly non-exploratory and incurious in the laboratory, but this may have signified little more than unusual interactions between their specialized behavioural adaptations and the arbitrary requirements for 'good performance' in the settings in which they were studied by Glickman and Sroges (1966) and by Parker (1974).

The scientific difficulties with the analysis of animal intelligence arise from the uniqueness of species and of the species-typical adaptive patterns within which behaviour patterns have their meanings. In its evolution, intelligence in any animal species would have evolved as part of a solution of a unique adaptational problem and must, therefore, have unique features. The inescapable inference from this uniqueness is that there cannot be a single trait we can call 'intelligence'. Rather, there would have to be a variety of intelligences (in the plural), although it might be appropriate to refer to any of these as generating or controlling intelligent behaviour.

This is not the place for a more detailed review of the uncertainties of definition and the failure to reach consensus. The failure has led to the occasional rejection of 'intelligence' as a distinguishing character for comparing animal species (Garcia 1975; Macphail 1982) and evolutionists may be sympathetic with the rejection. At any slice of time most species are, in a sense, equally well-adapted in every way to their respective niches, and the well documented difficulties in measuring differences among species in intelligence, defined by learning capacities or curiosity or other behaviour categories, show that intelligence cannot have been a simple and obvious factor in anagenetic evolution.

Despite the checkered history of efforts at definition, it is difficult to reject our intuitions about intelligence as a biological character. If we shift our ideas of intelligence from behaviour to experience (continuing to require objective criteria for describing behaviour as intelligent), we might emphasize how an animal knows reality rather than how it handles specific environmental challenges. We could then make an amalgamated neurological and behavioural statement about intelligence, which is both useful and intuitively acceptable. The statement would be as follows.

The brain's highest function is to create a 'real world' within which an animal can live and behave. Intelligence is the complexity of the 'reality' created by the brain. Consistent with an acceptable view of the brain's work (Jerison 1973), the formal definition of intelligence that is implied is almost morphological, although it has clear behavioural and physiological implications. It emphasizes general, statistical properties of large control systems, their capacity to process information, and their hierarchical organization, rather than specific brain structures and behaviour patterns. In my monograph and more formally in later publications (Jerison 1977, 1982b), I proposed this as a definition that relates intelligence to processing capacity. In addition to its suggestions about the comparative analysis of animal intelligence, the definition also enables us to use paleontological evidence on the evolution of brain size to interpret the evolutuion of intelligence in vertebrates.

A definition of biological intelligence

Definition. Biological intelligence in representative adults of a species is the behavioural consequence of available neural information processing capacity beyond that required for the control of general bodily functions

This is a definition for a species rather than an individual animal. It is also a definition involving the very general idea of processing capacity, suggesting only a division into two components, and obviously requiring more specification of 'general bodily functions'. Because of the importance of individual differences within the human species the first of these restriction to species rather than individuals should be explained more fully.

Biological intelligence must be studied 'above the species level' (Rensch 1959), because its between-species variance is apparently decoupled from within-species variance (cf. Lande 1979). Although the within- and between-species variance should be related, their decoupling may be thought of as analogous to the exhaustion of the heritability of a trait that occurs after consistent selection for that trait over many generations within a population (Falconer 1981). Individuals within the breeding population may then become relatively uniform with respect to the genetic determinants of an optimum value of a phenotypic trait. At the species level, if there are different optima in different species for a phenotypic trait under strong selection, each species may achieve comparable within-species uniformity (low heritability) correlated with its optimum. At the same time, the between-species variance for the genetic determinants can be high and reflect natural selection for the different values of the trait in the different environmental niches.

According to our definition, phenotypic 'intelligence' is linked to neural processing capacity, and there is good evidence that processing capacity can be estimated from brain size. However, the linkage, the covariation of processing capacity and brain size, is clear only for comparisons that involve different species. The decoupling of within-species from between-species covariation is most evident when processing capacity is measured morphologically as proportional to cortical surface area (Jerison 1982a,b, 1983). Decoupling is also suggested for behavioural processing capacity by the low correlation between brain weight and several behavioural measures in populations of mice bred for different brain sizes (Hahn *et al.* 1979). Our definition of intelligence can only be correct for comparisons among species.

It is possible to be precise about the component of processing capacity encumbered by 'general bodily functions' and to use inferences about the residual capacity as a basis for clearer ideas about intelligent behaviour.

The first component is considered in a theoretical analysis of expected brain/body relationships. The main inference from the residual capacity is that it may be used in many different ways in different species to support a variety of intelligences that have evolved.

Intelligence and brain organization

The brain's work is usually analysed in terms of its specialized projection and integration systems (Diamond 1979; Hubel 1979; Merzenich and Kaas 1980; Nauta and Feirtag 1979). The perspective on the brain for the analysis of intelligence is different, but it should not be misconstrued as opposing the conventional view. Rather, it is an extension of that view, and it is supported by recent analyses of the relation between projection and association systems, of the columnar organization of the brain, and of the relationship between processing capacity and measures of the surface and volume of the brain.

The perspective is probabilistic, or statistical, connecting the functions of brain systems to measures of gross brain size. Known specializations of various brain systems are not emphasized. Instead the goal is to identify 'parameters', which include (a) total neural processing capacity, (b) total numbers of processing units (cortical columns, etc.), (c) total numbers of neurons, and (d) average connectivity of neurons for a whole brain. Brain weight and cortical surface area are used as natural biological 'statistics' that estimate these parameters (Table 1, this paper; cf. Hofman 1982a,b; Jerison 1977, 1982a). Processing capacity is analysed into both structural

Table 1
Functions of brain size (E).

Dependent variable	Function	r	Reference[*]
1. Total cortical neurons (N)	$N = 1.04 \times 10^8 E^{0.66}$	0.96	2
2. Cortical volume (V)	$V = 0.5\,E - 1.9$	0.93	1
	$V = 0.31\,E^{1.09}$	0.99	1
3. Neuron density (total cortex)	$N/V = 10^8 E^{-0.32}$	−0.99	3
	$N/V = 2 \times 10^8 \times {}^{-0.31}$	−0.95	2
4. Glia/neuron ratio (G/N)	$G/N = 1.1 E^{0.31}$	0.99	4, 5
5. Length of dendrite tree (L)	$L = 0.11 E^{0.27}$	0.95	6
6. Chloride (Z; μmol/g)	$Z = 9.07 \log E + 30$	0.95	7
7. Acetylcholinesterase (Y; μmol/g/min)	$Y = 4.3 D^{-0.2}$	−0.996	7
8. Cortical surface (S)	$S = 3.74 E^{0.91}$	0.995	8

Note: correlation coefficients (r) are product moment correlations of the logarithms of E and the dependent variable for power functions. All units are in the centimetre-gram system.
    References: 1. Harman (1957); 2. Sharif (1953); 3. Tower (1954); 4. Friede and van Houten (1962); 5. Hawkins and Olszewski (1957); 6. Bok (1959); 7. Tower and Young (1973); 8. Jerison (1982a).

and functional components, but the parts are statistical rather than morphological; they are brain fractions rather than brain structures. The theory at its most elementary yet most fundamental level divides the brain into only two fractions, one of which controls the body, while the remaining (remainder or 'residual') fraction controls higher mental processes.

The fractions are not localized. Many parts of the brain that are known to map specialized projection systems may nevertheless contribute to both fractions of processing capacity described in our definition of biological intelligence. More specifically, the fraction that controls 'routine bodily functions' must involve neocortical as well as paleocortical, subcortical, and cerebellar structures. Processing of 'routine' visual functions, for example, occurs at many levels of the visual system, from the retine to several neocortical projection areas. Conversely, higher mental processes involve subcortical, paleocortical, and cerebellar structures as well as cortical systems of the brain (e.g., McCormick and Thompson 1984).

The relationship between capacities encumbered by routine bodily functions and those encumbered by 'intelligence' should be understood in the light of the relationship between projection and association areas in the brain. In at least two species, the tree shrew (*Tupaia glis*) and the bushbaby (*Galago senegalensis*), all of the neocortical surface has been accounted for by maps of projection systems, and there is no unmapped brain that can be labelled 'pure' association area (Diamond 1979). The integration of information from the various maps, which has traditionally been assigned to distinctly demarcated 'association areas', must now be recognized as embedded within the 'projection' areas. Particular areas of the brain can no longer be thought of as purely projection or purely association areas. They are information-processing areas in which other kinds of processing also occur along with specialized neural information 'projected' from sensory and motor surfaces.

To appreciate the importance of this new understanding of the organization of the brain, note that even if 'bodily functions' were controlled entirely by projection systems, and 'intelligence' were controlled entirely by association systems, these two components of processing capacity would not be localized in different parts of the brain. There may be different amounts of brain tissue devoted to each component, but the tissues would overlap morphologically. The primitive idea of localization of function, that neural systems that handle different kinds of information are located in different places, is not incompatible with this analysis. Specific control systems that contribute to the two components of processing in our definition may be localized as projection systems, but many different specific systems would be involved in each component.

Several recent developments in the neurosciences are especially important for the analysis of the work of the brain as a whole. There is a new appreciation of the number and diversity of cortical maps for particular 'projection' systems and the expansion of such systems far beyond their traditional 'areas' (Jones and Powell 1970; Bullier 1983). These do, indeed, cover the entire cerebral surface and include several polysensory regions that are presumably both 'projection' and 'association' areas in the old sense. There is, furthermore, a surprising structural uniformity in the mammalian neocortex (Rockel *et al.* 1980); local specializations apparently occur within the constraints of that uniformity. Finally, there is a growing consensus that the fundamental unit for processing information is a system of nerve cells rather than a single nerve cell. These systems in the mammalian brain are organized both structurally and functionally into columns of tissue, about 250 μm in diameter and about 1 mm deep, the full depth of the neocortex for cortical columns. Such columns have been most clearly identified in the visual and somatosensory cortex, but comparable systems may be present throughout the mammalian brain, subcortically as well as cortically (Hubel and Weisel 1979; Mountcastle 1978; Scheibel and Scheibel 1970; Szentagothai 1978). The columns are the modules of processing, analogous, perhaps, to chips in a modern computer.

From this perspective, the total processing capacity of a mammalian brain must be proportional to its total number of columnar modules. The modules could work in a variety of functional networks or systems, and thus be involved in various behavioural patterns. It will be useful to divide the behaviour patterns and the neural systems that support them, into the two categories suggested by the definition of intelligence. This can be done in a mathematically precise way, using the relationship between certain aspects of the brain's organization and brain size. At the same time, it is important to appreciate the way in which the projection systems work, their organization in living brains and the implication of that organization for the analysis of the evidence of projection systems as localized regions on fossil endocasts. That evidence may be used to suggest changing patterns of organization of the brain in evolving lineages.

Projection systems

The analysis of animal intelligence must be an analysis involving the evolution of the brain as a whole as an information processing system. Yet, in many ways, the classic picture of the brain as organized into identifiable sensory and motor projection systems remains valid. It is still possible to identify the major foci for processing visual information, auditory information, and so forth, and to locate these foci as areas in the brain. To a significant extent such areas are identifiable with respect to landmarks visible on the external surface. It is, therefore, possible to infer from the appearance of endocasts the extent to which certain kinds of information were processed in the brain of fossil mammals. It is possible to identify visual cortex on candid and felid endocasts and to make educated guesses about the expansion of frontal cortex in association with the evolution of their social behaviour (Radinsky 1973, 1979). The results of this kind of analysis are more or less what one would expect under the uniformitarian hypothesis. Specialized patterns of gyri and sulci recognizable in living species are also recognizable in their fossil ancestors at appropriate branchings of the fossil clades. The most important feature in such results is that there is no negative evidence. There is no suggestion of an evolutionary course contrary to inferences that would be made from living species and their interrelationships.

The most unusual localized system in the brain to which data from fossil endocasts have been applied is the human speech and language system. Two kinds of evidence have been examined that are relevant for the earliest appearance of specifically human behavioural capacities in the hominid record. Crucial evidence cannot be found in endocasts, of course, but if the external features identified in living human brains with language as a specialization were found at some specific stage in our fossil history, it would support the assumption that language in some form was present at that stage of human evolution. The necessary evidence would have to come from the asymmetry of the Sylvian fissure, which is the only morphological character potentially visible on an endocast that has been correlated with language in humans, but there could be supplementary evidence from the appearance of specific sulcal patterns in Broca's area of the left frontal lobe. In living primate brains there is slight asymmetry in the length and orientation of the Sylvian fissure noticeable in pongids and clearer asymmetry in the human brain (LeMay and Geschwind 1975; Yeni-Komshian and Benson 1976). Lemay (1976) and Holloway and de la Coste-Lareymondie (1982) have reported asymmetries for frontal and occipital 'petalia' (protrusions) in living and fossil hominids, but their functional significance is not clear at this time. There is no evidence of their relationship to language.

Lemay and Culebras (1972) reported asymmetry in the Sylvian fissure of the Chapelle-aux-Saints Neandertal endocast, based on drawings by Boule and Anthony (1911), which was comparable to that in living humans and correlated with language (see also, Geschwind and Levitsky 1968; Witelson and Pallie 1973; Wada *et al.* 1975). There are no other reports of asymmetries of the Sylvian fissure in fossils (cf. Holloway and de la Coste-Lareymondie, 1982).

Boule and Anthony presented their evidence in the form of pencilled 'Sylvian fissures' on the endocast, their guess of where the brain's Sylvian fissure had made its impression on the endocranial surface. I examined the endocast in Paris, their pencilled marks still clear, but thought their reconstruction of a Sylvian fissure imaginative. The evidence about language-related asymmetry from this or other endocasts is at best uncertain. It would, of course, be surprising if Neandertals had no linguistic abilities and no language areas in their brains; it is hard to imagine localized functional systems other than language systems that could require the mass of tissue available in their modern-sized brains. That may be the real evidence from Chapelle and other Neandertal endocasts. Their size is incompatible with anything other than the evolution of language (Jerison 1976b). However, the best evidence is from cultural artifacts. Neandertal artifacts that have been recovered (Marshack 1976) suggest a linguistic culture, which would require language areas to have evolved in this group of *Homo sapiens*.

Falk (1983) has recently analysed the fissural pattern of the frontal lobe of the famous *Homo habilis*, KNM ER 1470, which is about 2 million years old. She reports the presence of a convolutional pattern in the left hemisphere that could represent Broca's area, which would be evidence for the evolution of a speech and language area in this ancient hominid. The evidence would be stronger if a morphologically identifiable 'Broca's area' were evidence of language in living humans. It is not. A convolutional pattern of Broca's area is present in both hemispheres, but it is only in one hemisphere, normally the left, that speech is localized. In the opposite hemisphere there is also a set of convolutions that look like Broca's area, but these have not been related to language. The evidence from endocasts for the evolution of a 'Broca's area' may be no more that an expression of the unequivocal evidence that *Homo habilis*'s brain was significantly enlarged relative to the pongids and australopithecines. When brains evolve to larger size they necessarily evolve more extensive convolutional systems (Jerison 1982a, 1983), and the appearance of 'new' convolutions in the frontal lobe or elsewhere in ER 1470 was a necessary consequence of the enlargement of its brain.

There is little question that if language had appeared in *Homo habilis* it would involve a 'Broca's area' in the region identified by Falk, and to the extent that she demonstrated such an area she provided supporting evidence for specialization for speech in this ancestor of ours. It is not the strongest evidence, but it is probably the best that is available from the data on fissural patterns that can appear in endocasts.

This brief discussion should suggest the logical and scientific difficulties in inferring behavioural specializations from convolutional patterns visible in endocasts. The problems are with the specificity of the conclusions, and their dependence on unusually high correlations between localized functions and identifiable superficial structures in the brain. Perfect correlation would imply, for example, a perfectly localized visual area in the brain, mapped on gyri and sulci that are visible at the surface. There is, indeed, a localized area that should be called a visual area, but there is only a general, rather than point-to-point, relationship between finely analysed functions, and the exposed gyri and sulci, which are the only convolutions visible on an endocast. Visual cortex is one of the easiest to identify in mammalian brains, but the identification is not always easy in endocasts. It works in carnivores because their visual cortex is on the lateral surface, but in primates a large fraction is 'buried' in the longitudinal fissure, on the medial surface of each cerebral hemisphere, and is never visible in endocasts.

The most extensive analyses of data on the evolution of specialized functions in mammals as revealed in the fissural patterns of fossil endocasts have been reported by Radinsky (1979, etc.), and that is the most modern source for further discussion of this topic. Holloway (1975, 1981a,b, etc.) has presented comparable analyses for fossil hominid endocasts. Blumenberg (1983) presents an up-to-date bibliography on hominid endocasts, reviewing the speculations offered about them as evidence on the evolution of the human brain.

There are no narrowly localized functions of the brain that have been identified with intelligence or comparable higher mental abilities, although some 'higher' behavioural functions in mammals are affected by lesions localized in broader areas of the brain. Portions of the frontal and temporal lobes have most often been identified with such higher functions. Frontal lobe lesions typically produce deficits in performance on delayed response tests and double alternation tests, which are 'diagnostic' for a frontal lobe ablation syndrome. Masterton and Skeen (1972) showed that this effect could be reproduced without surgery, by taking advantage of the different evolutionary grades of insectivores (*Erinaceus europaeus,*) tree shrews (*Tupaia glis*), and primates (*Galago senegalensis*), in relative amounts of forebrain (and in encephalization, see below). These species differ in their ability to perform on the frontal lobe 'diagnostic' tasks just as if their natural differences in brain development had been generated surgically. There are probably other comparable correlations among other 'higher' functions and broadly localized structures, but in view of the overall correlation between the evolution of the parts of the brain and the whole brain most of these should also be recognized in the analysis of the evolution of brain size relative to the size of the body in mammals and the other vertebrate classes. The evidence for evolution of intelligence exemplified by the frontal lobe syndrome behaviour patterns is available in the history of the primates (Radinsky 1979) and it is also reflected in the history of encephalization as discussed in the next sections.

Brain organization and brain size

In my monograph and in a later review (Jerison 1973, 1977), I presented an elaborate analysis of the relationships between brain size and various other measures of the brain, which are summarized in Table 1. In terms of the discussion in the previous section, the easiest relationship to use is that between the surface area of the cerebral cortex and brain size, because it implies a simple relationship between processing capacity and brain size.

Other implications of Table 1 are worth noting, although they are not part of our main theme. Firstly, the data are remarkably orderly, and the orderliness anticipates the conclusion about the 'uniformity' of structure of the cerebral cortex (Rockel *et al.* 1980). Secondly, gross brain size is clearly a useful 'intervening variable' for the analysis of more detailed structure and function in the brain. Thirdly, it would be appropriate to take differences among species in brain size into account in research at both the cellular and molecular levels in comparative neurobiology. Finally, measures that have been proposed as indicating brain advancement and reorganization, such as the glia/neuron index and the extent of dendritic arborization (both probably related to connectivity), are related to brain size in a fairly straightforward way. The 'reorganization' indexed by such measures is indexed in the same way and to the same extent by gross brain size.

The important relationship for the analysis of intelligence is that of processing capacity to brain size. Function 8 in Table 1 estimates that relationship, because the number of columnar modules in the cortex must be proportional to the area of the cortical surfaces (Eccles 1979). Szentagothai (1978) described the diameter of a cortical column as constant across species, which implies a very simple relationship between the number of columns and the area of the cerebral surface. The other data in Table 1 would be more consistent with at least some increase in the diameter as well as the depth of a column as brain size increases. The expected increase in diameter would be small, however, and would not be noticed without special efforts to measure it. Whether the columns have the same diameter in different species or increase in diameter as brain size increases, there would be a proportionality between cortical surface area and processing capacity.

The connecting link between brain size and processing capacity provided by Function 8 in Table 1 is based on data by Brodmann (1913), and Elias and Schwartz (1970), who measured the brains of 48 species of mammals. I have described the analysis and its implications more completely elsewhere (Jerison 1982a,b, 1983). The impressively high correlation of 0.995 between the logarithms of brain size and cortical surface, a perfect correlation to two significant figures, almost suggests a functional rather than merely statistical relationship between these variables. If the cortical module is the unit of neural information processing, then total neural processing capacity is proportional to brain size. If the single neuron is taken as the unit of processing, functions 1–5 in Table 1 can be analysed as a group to support the same conclusion (Jerison 1973, 1977; see also, Hofman, 1982a,b).

This conclusion tells us what brain size means and why it is important. Brain size estimates ('measures' in a statistical sense) total neural information processing capacity. The next steps are to fractionate processing capacity into the two components required by the definition of biological intelligence, and then to examine the evolution of these components.

Factors in processing capacity

The analysis of the functions of the brain into two components was first proposed just over 100 years ago by Léonce-Pierre Manouvrier, Paul Broca's last student. Manouvrier (1883) wanted to analyse the brain 'mathematically', to determine a component of its size that was related to motor or somatic functions, and another component related to psychic functions. The idea that a two-component fractionation would be meaningful did not seem far-fetched to Manouvrier, nor should it to us. We can do the fractionation with an elementary regression analysis if we use body size as the covariate. When we regress brain size (= total processing capacity) against body size (= somatic demands on processing capacity) the residual should represent the remainder of processing capacity, or the 'psychic' component or factor, hence a two-component analysis.

An appropriate empirical regression analysis was first performed by von Bonin (1937), and has since been done by many others, most recently and significantly by Martin (1981) with the largest sample of mammalian brain/body data ever taken, over 300 species. The analysis has two parts, which have to be evaluated separately. There is first the classic allometric problem of determining either a fundamental or an empirical relationship among organ sizes, in this instance an allometric function that relates brain size to body size. Second, the residuals are analysed as measures of Manouvrier's psychic factor, or an 'encephalization' factor.

There is general agreement that the fundamental equation should have the form of the allometric equation

$$E_e = k\, P^\alpha \tag{1}$$

where E_e is the expected brain weight for a species of body weight P. The measurements of E and P are in grams, millilitres, or cubic centimetres, and the data are fitted to the logarithmic form of the equation

$$\log E_e = \log k + \alpha \log P. \tag{2}$$

Theoretical as well as quantitative issues arise in the determination of an appropriate value for α. In a regression model, the assumptions are that $\log E$ and $\log P$ are both Gaussian and that the residuals are random deviations from the regression. There are problems with both assumptions if we view the allometric function as in some way fundamental, because species with enlarged brains did not appear randomly in evolutionary history. If eqn (2) describes a primitive mammalian function, later changes in brain size in mammals as they evolved relative to this primitive function would probably have been limited to positive 'residuals'. For this reason, a regression analysis on living species could not determine the primitive function for mammals.

It is possible to take the regression analysis and its assumptions as heuristic, enabling us to determine eqn (2) objectively from data on actual brains and bodies. The computation is objective, but if we then treat the regression equation as fundamental we, in effect, assume that the deviations from 'primitive' allometric functions were independent of body size and averaged to an equal number of log units at all body sizes. In that case, the computation would lead to a non-primitive allometric function that would have the same value of α as the fundamental primitive function, and the difference between the computed value of k and the primitive value would be proportional to the average encephalization in the sample. This is the theoretical basis for using a regression analysis to determine the coefficients of eqn (2). The assumptions are then essentially of normality and homoscedasticity with respect to assumed average values of the coefficients for the population represented by the sample.

The ratio of brain size relative to expected brain size in a species, with the expectation determined by a regression equation, is an encephalization quotient, EQ. It is not always appreciated that EQ is, in fact, a residual from a brain/body regression analysis and it is worth a brief digression to demonstrate the identity. Note, first, that eqn (2) may be recast as a regression equation by adding an error term for any species i. The regression equation is then

$$\log E_i = \log k + \alpha \log P + \log e_i \tag{3}$$

Species i has a brain weight E_i, which deviates from the expected brain weight by the residual 'error' factor e_i. The expected brain size for body size P is given by eqn (2), and the encephalization quotient is defined as

$$EQ = E_i / E_e$$
$$\log EQ = \log E_i - \log E_e. \tag{4}$$

Subtracting eqn (2) from eqn (3), we have

$$\log EQ = \log e_i$$

and

$$EQ = e_i \tag{5}$$

for species i. Thus, the encephalization quotient is exactly equal to the antilogarithm of the residual.

The components of brain size that might be called 'regression' and 'residual' are more frequently described as 'allometric' and 'encephalization' components. The allometric component implies an underlying relationship between brain and body size due to structural constraints in 'average' animals (Gould 1966, 1975). The encephalization component is also a structural measure, and can be used to indicate how much a species deviates from the allometric norm in the amount of neural machinery in its head.

The allometric factor in brain size

The strictly evolutionary questions about allometry are 'uniformitarian'. Are the coefficients α and k of the allometric function, eqn (1), the same now as they were early in the evolutionary history of the taxa to which they refer? Is encephalization evident in fossil species consistent with the present diversity of species? Is it reasonable to interpret the allometric coefficients in the same way in living species as in primitive species? We assume positive answers under the uniformitarian hypothesis and are, therefore, concerned with the results of allometric analysis in living species.

The empirical results of brain/body allometry in living species can be summed up fairly easily (Jerison 1983; Martin 1983). When large numbers of species of mammals are sampled, and if the species are heterogeneous in body size and adaptive niche, studies prior to 1980 reported exponents of about 2/3. These begin with von Bonin's (1937) analysis, the first that used a regression model to determine the exponent. I obtained the same result (Jerison 1973), using data from Crile and Quiring (1940) on 100 species of mammals. Since 1980, such studies have more frequently reported exponents of about 3/4, following Martin's (1981) review of mammalian data. Studies, both before and after 1980, when sampling mammalian species that are similar taxonomically, i.e., from the same genus or the same family, report exponents of between about 0.2 and 0.6, with typical values of about 1/3. Within-species, as in comparisons of breeds of dogs (Bronson 1979), the exponent is between 0.2 and 0.3, but within other species, such as rhesus monkeys and humans, the exponent may not be significantly different from zero; that is, brain size may be functionally independent of body size (Jerison 1979b, 1983).

There is one especially interesting datum in this literature, from work on breeding mice for small and large brain size (Roderick *et al.* 1976). Although their data are on the small sample of three breeds of mice, their exponent, $\alpha = 0.77$, is the only within-species exponent that I have ever seen that was greater than about 0.6. Lande (1979) explained this exponent in an intriguing evolutionary analysis of the allometric function.

Lande pointed out that the allometric exponent can be interpreted in evolutionary population-genetic terms and its value may then reflect selection for brain size or for body size. The theoretical equations assume that selection pressures are entirely on brain size or entirely on body size. In the breeding experiment, selection was entirely for brain size, of course. In comparisons among species of mice, as natural populations, one would assume selection for body size.

The variables for the model are the heritabilities h^2 of brain and body size, the genetic correlation r between these sizes, and the variances s^2 of these sizes as measured on a scale of natural logarithms; the variances are approximately the same as coefficients of variation. We may define

$$Q = \{(h_{\text{brain}})\,(s_{\text{brain}})\} \,/\, \{(h_{\text{body}})\,(s_{\text{body}})\}.$$

Lande showed that under selection for brain size

$$\alpha = Q\,/\,r,$$

while under selection for body size

$$\alpha = r\,Q$$

The genetic correlation between brain and body is normally positive, and between zero and 1; since $Q < 1$ for brain and body measures, it is evident that allometric exponents will be higher for selection for brain size than for body size. Lande's model can account for the distinctions among the various samples mentioned earlier with respect to exponents.

Lande showed that, with reasonable assumptions about heritability, variance, and genetic correlation, we should expect the 0.77 allometric coefficient in mice bred for brain size (selected for brain size) if natural populations (assumed to be selected only for body size) yielded within-species exponents of 0.36, which are typical exponents for comparisons among natural populations of mice.

Lande's model is one modern solution of Manouvrier's problem. Selection for brain size would result in the development of a psychic factor, as it were, and selection for body size would result in the development of a somatic factor. In terms of the regression model, selection for brain size should affect encephalization and selection for body size should affect the allometric factor, to the extent that it is independent of encephalization.

The allometric function has always been interpreted in at least two other ways: as related to information processing and as related to bioenergetics. Both relationships were suggested by Snell (1891) in the first published theoretical analysis to support $\alpha = 2/3$ as a fundamental constant in the equation, a value he believed to be consistent with both interpretations. Kleiber's (1949) and Brody's (1945) reports that basal metabolic rate 'scales' with body weight according to $\alpha = 3/4$ made it appear that a 2/3 exponent for brain/body relationships could only reflect information processing constraints. When exponents of 2/3 were reported in regression analyses of brain/body data in living and fossil species (von Bonin 1937; Jerison 1961) these were taken as support for an informational analysis, specifically as reflecting the mapping of body surfaces in the brain (Jerison 1977).

Martin's (1981) review of the brain/body exponent, in which he reported that the empirical mammalian exponent is 3/4 rather than 2/3 for a sample of 309 species of mammals (see also Armstrong 1982; Bauchot 1978; Hofman 1982a,b) may be interpreted as providing evidence for an energetic constraint on brain size. There is, in fact, no inconsistency between scaling both brain size and basal metabolic rate against body size according to an empirical mammalian $\alpha = 3/4$, and choosing other values of α for analysing the role of brain size in information-processing.

Martin (1983) implied that $\alpha = 3/4$ is an upper boundary value for samples of species that are similar in energetic requirements as well as broad morphological features. In that case, the relation between brain size and energetics should be maintained to the extent that there is geometric similarity in mammals. Hemmingsen (1960) has demonstrated such similarity by showing that $\alpha = 2/3$ for the regression of body surface on body volume. An analysis of brain/body relations in a very broad sample of mammals could, therefore, reflect the same metabolic constraint on brain size as on the size of other organs.

In the case of the brain, if equally encephalized species scale at $\alpha = 2/3$, or less with respect to information processing, and if encephalization is correlated with body size (cf. Bauchot 1972, 1978; Rensch 1959), larger species would tend to be more encephalized than smaller species. Such an evolutionary trend could generate empirical exponents significantly greater than 2/3, and the observed 3/4 for the mammals as a whole could represent a kind of maximum, an energetic limitation on the average amount of encephalization that could evolve. (This would, incidentally, agree with the empirical exponent in laboratory mice that were bred for brain size, as mentioned earlier.) The evolutionary trend is an expected one, because the earliest mammals were uniformly small-bodied and trends toward larger bodies are common in evolution (common enough to have been named 'Cope's law'). Even if it occurred randomly among evolving mammalian species, encephalization would be expected to be correlated with body size, because more large-bodied species appeared later in evolution than earlier in evolution. There is, therefore, no necessary causal inference from a correlation of encephalization and body size. These can be independent, each correlated with time as a hidden covariate.

Nobody understands why $\alpha = 3/4$ works empirically as the 'scaling factor' for energetics and body size. Whatever the reason, it will obviously affect many morpohological traits, including brain size, in which bio-energetic constraints are important. The reasoning in the previous paragraph suggests that it is premature to use an exponent of 3/4 as the basic mammalian allometric function for all analyses of encephalization. Allometric functions are normative, used to indicate a kind of equality with respect to organ systems in animals of different body size. Species lying on the same allometric curve are equal to one another, as it were. The brain as a functional system can be analysed at many levels. At one level, it is an energy-exchange system and, accepting Martin's analysis, species on the same allometric curve defined by $\alpha = 3/4$ are comparable to one another in that respect. It is also an information-processing system and, for theoretical reasons, the normative allometric function should be significantly lower when it reflects equal amounts of processing capacity.

Some theory on allometry and encephalization

In Jerison (1977), I discuss a theoretical basis for $\alpha = 2/3$ as a maximum because of the representation of information about the external world as maps in the brain. This exponent does not mean that the body's surfaces in different species must contribute in the same way to brain size; they probably do not (Armstrong 1982). The exponent relates two-dimensional maps in the brain to three-dimensional bodies that are used to estimate the areas of the maps. The 2/3 exponent is required for balancing a theoretical 'allometric' function dimensionally. A maximum empirical $\alpha = 3/4$ or so is consistent with the mapping function and it is worth another digression to see why.

The final equation in my theory of encephalization is:

$$E = (0.1)\,(m)\,(p^{2/3}) + A \qquad (6)$$

which is very similar to eqn (1) in structure: $k = 0.1m$ and $\alpha = 2/3$ when $A = 0$. The terms m and A indicate different aspects of encephalization: an amplification of the mappings in the brain and an increment in 'associative' activity, respectively. (I do not support the idea of one-dimensional encephalization, measurable by EQ, except as a convenience.) The m-term is a non-dimensional amplification factor. It multiplies the area of the surface represented by $p^{2/3}$ (which represents about 10 per cent of the surface area of the body; Hemmingsen 1960), to be proportional to the areas of the summed maps in the brain. When m is the same in different species the species are equivalent in the amount of mapping of the external world that goes on in their brains. When species differ, it should reflect the presence of either more maps or more detailed maps in their brains. Primates, for example, have more maps of the retina projected on the visual systems of their brains than other orders of mammals; less visual orders such as rodents, have fewer or less extensive maps (Merzenich and Kaas 1980). In this sense, there is more amplification of visual information in primates than in rodents. To the extent that all mapped systems in the brain contribute to brain size, the factor m reflects their summed contribution, compared to other species. Other points: $EQ = m$ where $A = 0$, and the number 0.1 is a length in centimetres (the 1/3-power of body weight or volume) to balance the equation dimensionally. I suggested that it could represent the depth of a modular column which would also be the thickness of the neocortex.

Martin (pers. comm.) pointed out that I simplified by treating the depth term as constant at about 1 mm, and that if I accepted the small relationship known to exist between the thickness of the neocortex and brain size, I would find that 3/4 is a correct theoretical maximum for empirical allometric exponents. He is right. An allometric function with $\alpha = 1/6$ or so relates neocortical thickness to brain size (Harman 1957; Jerison 1982a), and if we take the traditional 2/3 as the brain/body allometric exponent, we find that the dimensionality of the depth term would have to be partitioned into one part for body weight, $(2/3)(1/6) = 1/9$, and the remainder, $3/9 - 1/9 = 2/9$, still attached to the depth term. The effect would be to have the body weight term contribute at $p^{2/3}$ for its role in determining the area of the maps and $p^{1/9}$ for the depth of the maps, $(p^{2/3})(p^{1/9}) = p^{7/9}$. Thus, $\alpha = 0.78$, about 3/4, is the maximum empirical exponent to be expected in a sample of species that are equally encephalized with respect to information processing.

When $A>0$, the graph of eqn (6) is a concave curve above the regression line (in which $A = 0$), approaching it as a limit on log-log co-ordinates. A straight line fitted to the same data has lower values of α when A is large than when it is small and this is part of my explanation for the different values of α obtained in different kinds of samples. To illustrate, in eight species of Cercopithecus monkeys, fitting eqn (2) to their brain/body data resulted in $\alpha = 0.32$, $k = 4.6$. Fitting the same data with eqn (6) resulted in $m = 1.0$, $A = 37.5$ g and this would be my preferred way to describe species that are equally encephalized, but differ in body weight. The important point is that the graphs of these two equations are indistinguishable for the range of body sizes represented by the eight species in the sample (Jerison 1982b).

The exercise with the Cercopithecus monkeys suggests a theoretically acceptable way to measure encephalization with an EQ. To compare groups of species with respect to information processing capacity one could determine encephalization quotients for the lowest relevant taxonomic category, rather than the highest as suggested by Martin (1983). This is essentially Dubois's (1897) procedure, provided with better statistical underpinnings, of determining allometric functions for each group, averaging the exponents, and then forcing the fits of all the groups to be compared to an allometric function with the average α. (α must be the same if the groups are to be compared as residuals from the same regression.) The resulting encephalization quotients (residuals) would be measures that compare the groups with respect to adjusted processing capacity by combining the effect of factors m and A of the theory, which are informational constraints on the empirical exponents. The method should work as long as there is no more than about a five-fold difference in body sizes, a range of about 3/4 of a log-unit.

Empirical allometric exponents often reflect *ad hoc* combinations of constraints rather than fundamental relationships, and it is important to appreciate this when using allometric functions. Szarski (1980) reviews many of the constraints that should be expected and considers the empirical exponents to be best explained by the interaction of the constraints, a position that seems to me almost self-evident. It is the only way to explain the variety of empirical exponents obtained reliably in different kinds of sampling.

The allometric exponent of 3/4 fits the data for mammals as a class, but as the sampling is restricted to lower taxa the exponent becomes smaller and smaller, as mentioned earlier. The effect is predicted by eqn (6) if the groups have a significant contribution of the 'associative' factor. One problem with the metabolic interpretation of brain/body allometry is that Kleiber (1947) found the same 3/4 exponent for metabolic-rate/body-size allometry within-species as well as between-species. The metabolic constraint on brain-body relations apparently operates only on evolution very much above the species level. Brain/body allometry at the generic and species level is apparently governed by other factors.

Martin's analysis as presented in detail (Martin 1983) is appealing. It focuses attention on the relationship between energetics and brain size, which leads directly to concern with comparisons among growth rates within different species and the place of ecological factors in present variations among species, issues discussed at length by Martin (1983) and others (Armstrong 1982; Sacher 1976; Sacher and Staffeldt 1974; Szarski 1980). It is helpful to be reminded that the brain is an organ of the body, and has to be built and work under bodily constraints, while recognizing that its function, in the biological sense, is as an organ that handles information.

Body size and allometry in fossil vertebrates

Allometric functions have been fitted to many fossil data (Hopson 1979; Jerison 1973, 1979a), but in no instance was the sampling appropriate for crucial tests of exact values of exponents. The evidence is consistent with allometric exponents having been approximately the same throughout evolutionary history. There is no reason to assume that the scaling of brain size against body size has changed. The easiest way to recognize this evidence is by drawing convex polygons about sets of data, rather than by fitting regression lines. Such polygons are of approximately the same shape and are oriented in brain/body 'space' (the log-log co-ordinate system for brain/body data) at angles consistent with the values of the exponents of allometric functions for living species. More precise analyses require reasonably accurate estimates of brain size and body size.

The estimation of body size has been a controversial problem in brain/body allometry in fossils mammals. My methods (Jerison 1973) included a kind of multiple regression analysis in which body weight in fossils was estimated from the cube of the body length of skeletal reconstructions and from a second binary variable, to which I assigned values of 0 or 1 as I judged the animal to be 'average' or 'robust' in habitus. Radinsky (1978) properly criticized the subjectiveness of this procedure, but 'corrected' it by ignoring robustness and using the bivariate regression of body weight on body length in living species of a given taxonomic order for the estimation of the weights of fossils of that order. He used an averaged regression for fossils of extinct orders. The real problem with this approach is the uniformitarian hypothesis: is the present array of body shapes in mammals an adequate guide to the shapes of fossil species? In some ways it is, but in crucial ways it appears not to be.

Fossils were probably sufficiently similar to living species in shape to justify the reconstruction of a whole animal from a skeleton, adding 'musculature' and 'skin' consistent with the thickness of the bones, the size of protuberances on which muscles insert, and so on. In living species this procedure results in infrequent errors; the camel's hump and the elephant's trunk would be missed, but most living animals can be reconstructed reasonably well in this way. There is a branch of forensic anthropology that makes its way by correlating human facial features with skulls to enable police to identify the skulls with individuals who have disappeared; however, the procedure with fossil animals is difficult and tedious. Weight/length regression analysis based on the data of living species, a much easier procedure, would be appropriate if the shapes of fossils were correlated with length in the same way as in living species.

Although such a correlation is generally reasonable, uniformitarianism appears to fail in one respect, in that early in the evolution of larger mammals, at the beginning of the Tertiary period, there appear to have been many more small, yet robustly built species than there are today. This is evident in even a cursory look at the skeletons of many early fossils, especially of archaic orders of the Paleocene and Eocene epochs. The evidence is usually in the size and shape of the long bones, especially the humerus and femur, which sometimes look like isometrically scaled down bones of large bears in animals with body lengths of wolves. Animals built in this way are likely to be unusually heavy compared to living species of the same body length and underestimation of body size will be frequent.

One can test this question by reconstructing whole animals from skeletal information and comparing the volume of the reconstruction with the estimated volume according to the regression equation. It was this procedure that led me to use the subjective binary variable in my analysis (Jerison 1973, p. 239). More accurate regression analyses may be possible using data other than body length in living species. Alexander (1981) has reviewed some of the structural constraints on body size. With his associates (Alexander et al. 1979), he has published analyses using the regression of femoral diameter against body weight in mammals, which suggests that appropriate multiple regression equations using such measures could replace my approach in a satisfactory and objective way.

Since good data on brain size are available from fossil endocasts, it should be possible to estimate allometric functions for reasonably large samples of species, and if the evolution of encephalization is to be studied with encephalization quotients based on appropriate allometric functions it is clearly important to develop methods of estimating body size. Even better estimation is possible now by using computer graphics to reconstruct bodies from skeletal materials, in much the same way as this has been done laboriously to make museum displays. The volume of the 'body' can then be 'measured' from the data used to generate the computer display.

I remain convinced that my estimations of body size in fossil mammals, despite their subjective element, are the best published estimates at this time. In the light of Radinsky's criticisms, however, the consensus necessary to support subjective approaches is obviously not present. However, there is no reason to accept either the subjective element or the inadequate bivariate statistical model. Objective multivariate analyses can control 'robustness' statistically, independently of body length. Body size is, after all, a simple conceptual problem, comparatively speaking, and consensus should be forthcoming from very little additional analysis. Almost the reverse may be said of brain size. Here the measurements have been easy, with consensus the rule rather than the exception, but the conceptual problems have been difficult indeed.

The evolution of encephalization: the classes of vertebrates

In addition to Martin's (1981) compilation of data on species of mammals, there are now many more published compendia of brain and body weights in all the living vertebrate classes. Minimum convex polygons about the log-log data of each class are the simplest and clearest evidence of encephalization. Adding fossil data clarifies instead of confusing this picture.

Mammalian and avian polygons are almost congruent, differing primarily in the absence of birds weighing more than about 100 kg (Jerison 1985; Mlikovsky 1977). Polygons containing the data of bony fish (Ridet 1973), amphibians (Thireau 1975), and reptiles (Platel 1979) lie below and do not overlap the first pair of polygons. These three lower polygons also form an almost congruent system, with a trailing amphibian polygon that includes the smallest vertebrate species that have been measured. Representing five vertebrate classes, the two sets of polygons are orientated at similar angles in brain body space, displaced vertically from one another. Since birds and mammals evolved from reptilian ancestors, the graph is interpreted as indicating an evolutionary advance in grade measured by the displacement of the upper composite polygon from the lower one. The increase in brain size relative to body size in the advance from 'lower' to 'higher' vertebrates is the clearest and most dramatic of many examples of encephalization.

Data from the cartilaginous fish and the jawless fish (Ebbesson and Northcutt 1976; Northcutt 1981) complete the picture of the evolution of encephalization. The elasmobranches form a polygon that lies between those for lower and higher vertebrates, and overlaps both. Some sting rays are at a mammalian or avian grade of encephalization, and that favourite 'primitive generalized vertebrate' of the comparative anatomy class, the dogfish (*Squalus*), is more encephalized than any of the 'lower vertebrates'; its brain/body datum almost reaches the lower bound of the higher vertebrate polygons. Species of cartilaginous fish may have been the first vertebrates to 'experiment' with enlarged brains. Agnathan data (*Petromyzon*) fall well below the 'lower' vertebrate polygon; the transition to the gnathostome grade may have been the first advance in encephalization in vertebrates.

These facts in living species are the framework for interpreting the fossil record, which was the central concern of my monograph (Jerison 1973). A new and still unpublished datum seems to confirm the inference about the place of the elasmobranchs in the history of encephalization. R. Zangerl (pers. comm.) described a small Permian shark in which the head region was relatively uncrushed; its endocranium was comparable in size and shape to that of its living, highly encephalized relatives (e.g., comparable to *Squalus*). The confirmation was with a single specimen, but a single case is all that is needed for the analysis. The polygons define regions in 'brain/body space' for grades of encephalization. If the fossil shark lies above the lower vertebrate region it was not a 'lower vertebrate' from this perspective.

The advance in grade from the agnathans to gnathostome fish evident in data on living species is unlikely to be confirmed by fossils as an anagenetic change. A history of the events, which occurred over 400 million years ago, is recorded in the crania of the many fossils that are known from both vertebrate groups, but the evidence of the fossil agnathan brain, although impressed as beautiful patterns on the endocranium (Stensiö 1963), is too poor to permit a judgment about brain size (Jerison 1973). Living lower vertebrates could represent the conservation of a primitive vertebrate condition, while living agnathans could be degenerate descendants of species that were at a 'normal' primitive grade, as far as the fossil evidence is concerned.

Research and analysis of the fossil evidence since the publication of my monograph has provided new information on transitional grades between vertebrate classes. In general, the differentiation between lower and higher vertebrates has been confirmed, but apparently there were 'experiments' with enlarged brains in the ostrich-like dinosaurs (Ornithomimidae; see Russell 1972); some of the mammal-like reptiles may also have had enlarged brains (Kemp 1979; cf. Quiroga 1980). The orientation of the polygons for lower and higher vertebrates when fossil data are added is the same as that for living species – they lie at a slope of about 2/3. The vertical displacement is also similar, although the transitional or experimental forms are easily identifiable on such graphs as falling near margins of polygons.

Sampling the fossil data on reptiles extends the range of sizes to make the reptilian polygon comparable to that of living mammals in every respect. Hopson (1977, 1979) has confirmed my conclusion that the dinosaurs, as a group, were normal reptiles in relative brain size, suggesting in addition that the predatory dinosaurs were more encephalized than their prey, the large herbivores. Dinosaurs did not become extinct because of their small brains.

I have reported that *Archaeopteryx* was a transitional form in encephalization, lying below the avian polygon. This interpretation is confirmed by a new preparation of the endocast (Whetstone 1983). A new specimen described by Wellnhofer (1974) appears to be brainier – its datum is within the polygon of living birds (Hopson 1977). New data on later birds (Mlikovsky 1980) indicate that the Miocene hawks had reached the present grade of encephalization.

Encephalization is a topic in anagenetic evolution 'above the species level' (cf. Rensch 1959), clearest at the highest taxonomic levels and murkier as it deals with lower taxonomic levels. The problem is formulated as a search for information about the likelihood that a particular grade of encephalization will have appeared in a particular taxon at a particular time. Issues on rates of evolution, such as punctuated equilibria versus gradualism, can be addressed, although the fossil record of the brain and the finer analysis of encephalization are not yet good enough to be anything more than suggestive about the answers. On the evidence, it seems that when there have been advances in grade in a broad spectrum of species, e.g., in the prosimians, the advance occurred over at least several million years, but when an appropriate grade was achieved there was no further advance. The prosimians appeared to have reached their present grade of encephalization by the end of the Oligocene (Gurche 1982; Jerison 1973, 1979a; Radinsky 1977).

Steady states such as the Neogene stasis in prosimian encephalization are the rule (Jerison 1983). The picture is, therefore, more consistent with a punctuational than with a gradualist model. The steady states presumably represent the stabilization of niches and the consequent optimization of those behavioural adaptations in which encephalization is involved. However, the time-grain for the analysis is so coarse that even in the 'rapid' encephalization of the hominids it is impossible to assert that a punctuational model is superior (cf. Hofman 1983).

Encephalization in fossil mammals

Unlike the estimation of body size discussed earlier, the estimation of brain size in birds and mammals, living or fossil, is not a serious problem. Endocasts and brains are very similar in volume in these classes, differing by no more than 10 per cent or so. Therefore, it is possible to analyse encephalization in fossil species in some detail. Fossil birds, unfortunately, continue to be rare finds for bodies and brains (cf. Mlikovsky 1980) and the general outline of their encephalization as I described it in my monograph remains as valid, but as incomplete now as it was a decade ago. The story in mammals is different.

Writing on the basis of the small samples available a decade ago, I was able to discover trends of encephalization that seemed excitingly like those that a committed Darwinian would expect in the material (Jerison 1973). I found 'leap-frogging' evolution of encephalization (Simpson 1953), an interaction between the major orders of ungulate herbivores and the orders of carnivores that preyed on them, which suggested that improvements in defense interacted with improvements in attack, as in the recent history of warfare and international relations (Dawkins and Krebs 1979). I also found slower evolution in Neotropical herbivores that were not under the predacious pressures of their remote Holarctic cousins.

I still believe that these trends are correctly pictured in my monograph, but the most critical analyst of this material (Radinsky 1978, 1979, 1981) disagrees. If one does not take body shape into account many of the neat relations that I saw in the data may dissolve as too vague to analyse. Radinsky (1978) does not accept my subjective designations of fossils as 'average' or 'robust' and without consensus about the evidence of one's perceptions of fossil skeletons it is impossible to maintain those designations as components of a regression analysis. The problem was discussed earlier and it has a simple solution: estimate body size on the basis of more skeletal information than simple body length. The information is available in the long bones of fossils (Alexander *et al.* 1979), and its analysis should provide a statistical measurement of robustness as one of the variates in a multiple regression.

The analysis has not yet been performed. My conjecture is that it would not show the clear advantage of predators over prey in relative brain size that I found (see also, Jerison 1976a). The reason is that a slice in time for the present, in which living carnivores are compared with living hoofed mammals with respect to encephalization, indicates that the groups are about equal in encephalization (Jerison 1973, Appendix II). If no difference could be detected in that larger sample, it is obviously unlikely to be detectable in the more haphazard and smaller samples of fossils; if a difference were to appear in fossil samples from closely related fauna one would conjecture that it was an artifact of sampling rather than evidence of a changed relationship in fossil groups as compared to recent groups. As in many analyses, the time grain is too course to substantiate leap-frogging in encephalization on the basis of the fossil evidence.

The scientific question should really be put in the opposite way. Our interest is less in the evidence of the limited, and inevitably biased, data of a few fossil brains and bodies, than in the reality of the phenomena in which encephalization may have interacted with ecological niche and behavioural adaptations of prey and predator. The 'null hypothesis' should be that there is a difference between prey and predator, and we should look at the data for evidence of no difference. In other words, with such weak data we do better to treat them as a source of falsification of hypotheses rather than of verification. In that case we may assume the leap-frogging to have occurred and then ask whether the fossil data controvert the assumption. The answer is no.

A requirement for more encephalization in predators than in prey is easy to understand (cf. Dawkins and Krebs 1979) and is independent of a particular set of data. The predator's problem is more diffuse than the prey's. It must adjust to the various defensive strategies available to prey species, and successful attack can depend on surprise, on the unexpected from the defender's vantage point. To deploy one's energies in surprising ways is clearly more difficult than to adopt a successful defensive strategy, which may be the immobilized strategy of the besieged. The difference is exactly the sort that would be reflected in some increase in processing capacity, but increments have to be enormous to appear in measures of encephalization and few adaptations actually require so much special processing capacity.

Hopson (1979) sees evidence for a difference between predatory and prey dinosaurs in their encephalization, and perhaps the absolute requirements for neural processing capacity may be great enough that they will appear in an analysis of relative brain size for small brained species as opposed to large brained species, i.e., in reptiles as opposed to mammals.

A stronger argument may be possible with regard to the related ecological issues in the fossil Neotropical mammals of South America. I had found these to be less encephalized than their Holarctic counterparts (Jerison 1973, Ch. 14). A Darwinian explanation is straightforward: there were no 'progressive' predator species among the contemporaries of the Neotropical ungulates during their adaptive radiation. Elegant though it was, my discovery was not confirmed by Radinsky (1981), possibly because he did not take robustness in body size into account. To resolve the problem a better analysis of body size in fossils is necessary and, as indicated earlier, it should be easy to perform.

These unresolved problems have repercussions for our understanding of the diversity of brains in living species. The issues are ecological. What relevance is there in brain size, encephalization, or information processing capacity for the likelihood that a species will be able to function in its niche? To what extent do these matters contribute to fitness? These were Radinsky's legitimate questions. They should and they can be answered. In living species the same kinds of questions have been posed. The relationship of herbivore to carnivore, of fructivore to insectivore, of hunter to scavenger, of predator to prey are at the core of the problem. Important relationships almost certainly exist, which may be strong enough to appear in the crude analysis of behavioural capacity, available from the analysis of brain/body data. Clutton-Brock and Harvey (1980), Eisenberg (1981), and Harvey et al. (1980) have performed various analyses of the correlation between encephalization and ecological niche; they are reviewed critically and cogently by Martin (1983).

Large-brained mammals

Among all the orders of mammals three stand out in brain size and encephalization. These are the pachyderms, the cetaceans, and most remarkably, the primates. The African and Indian elephants are all that remain among living species to represent the once great order of Proboscidea. Their brains are known morphologically, but not physiologically. They are simply too hard to get at to study while an elephant is still alive. The brain is large, even in relation to an elephant's body size, but it is about as large as expected according to Martin's $\alpha = 3/4$, and only slightly larger than expected ($EQ = 1.5$ or so) according to $\alpha = 2/3$. The intelligence of elephants is proverbial, but the scientific evidence for it (or against it) is not worth discussing (cf. Rensch 1959).

The cetaceans include the one non-primate species that is comparable to humans in encephalization, the bottle-nose dolphin *Tursiops truncatus*, and its close but larger relative, the killer whale, *Orcinus*, with a much larger brain and body, but a lower encephalization quotient, in spite of its capacity to learn and perform marvelous tricks. Cognitive capacities of dolphins are more than proverbial. Herman (1980) was impressed enough by them to suggest that they were 'convergent' in an evolutionary sense with that of the great apes. My own view is that Herman may be conservative on this score (Jerison 1982b, 1985).

The primates are in a class by themselves, of course, because one of their number is *Homo sapiens*, a species that includes the reader and the writer. A sense of the present status of research on the evolution of the primate brain is available in Armstrong and Falk (1982), and an enormous experimental, field-observational, and theoretical literature is extending our perspectives, with major contributions reflecting the development of sociobiology (see Dawkins and Krebs 1979; Lumsden and Wilson 1981). The analysis of encephalization based on fossil data is fairly simple compared to much other work going on now. Three conclusions are evident to me in the history of encephalization in primates as relevant for the human condition. Firstly, primates have always been a brainy order, perhaps doing with their brains what many other species did by morphological specializations. Secondly, the evolution of encephalization in the primates followed rather than preceded or even accompanied the enlargement of the brain. Washburn (1978) has pointed this out as a feature of hominid evolution, but it appears to have been true for prosimians and simians as well (Jerison 1979a). Finally, the specialized advances in hominid encephalization are surely correlated with one unusual cognitive adaptation, a new or at least expanded way of knowing reality, namely, the language 'sense' (Jerison 1973, 1983).

'Knowing' is cognition and I have asserted the not unfamiliar doctrine that language is a cognitive system, that language and thought are intimately related. Although the doctrine is familiar, its implications for evolutionary neurobiology may not be fully appreciated. Cognitive systems are the systems that construct our realities. They are sensory/perceptually integrated with motor systems, which account for much of the mass of the brain in more encephalized as opposed to less encephalized vertebrate species. That human language is also a communication system is secondary from this perspective, since communication systems in most vertebrates are handled with a relatively small investment in their neural control apparatus (MacLean 1970). Our enlarged brain is accounted for primarily by its role in supporting our strange cognitive adaptation that is language.

The role of language in creating reality is well appreciated. The definition of intelligence as identical with encephalization implies that realities may be created in at least slightly different ways in different species, since encephalization can occur in many different ways. Language is the peculiarly human adaptation correlated with our extraordinary grade of encephalization. It is, in a sense, the measure of the complexity of the reality that *Homo sapiens* creates. That creation has been the result of one evolutionary path in the evolution of one organ system in one mammalian lineage. Where there has been encephalization comparable to that in humans in a species, there is a challenge to try to discover the analogue for 'language' in that species. The analogy would not necessarily be a communication system. It should be a 'cognitive system', based on sensory and motor information that contributes to the reality constructed by the brain of the other species.

There is only one species, *Tursiops*, that meets the criterion of an essentially human grade of encephalization. The large-brained bottle-nosed dolphins, like chimpanzees, can be trained to use a human syntax for communication with their trainers (Herman 1980), and that in itself is a measure of the great 'intelligence' of these two distantly-related species. It would be even more of an achievement to discover the natural 'language' in the sense of the 'reality', rather than mere communication, of exotic mammals in their unusual environments, to let ourselves be trained (as we were trained by bees to learn their 'language') by the animals in their knowledge of reality.

The brain evolved in many directions and its work should be appreciated with respect to all of those directions. Diversity has been the rule in the outcome of this evolution and the most interesting conclusion that can be offered is that this diversity must be even more dramatic for the very encephalized than it is for small-brained species, that the evolution of intelligence in relation to the evolution of very large brains has probably been an evolution of many kinds of intelligence. We know encephalization in other species and we know directly the correlate of encephalization in one species, our own, the intelligence in which language has so strange a role. Comparably exotic intelligences remain to be discovered in at least some other species.

References

Alexander, R. McN. (1981). Factors of safety in the structure of animals. *Sci. Prog. (Oxford)*, **67**, 109–130.

—— Jayes, A. S., Maloiy, G. M. O., and Wathuta, E. M. (1979). Allometry of the limb bones of mammals from shrews (*Sorex*) to elephant (*Loxodonta*). *J. Zool. Lond.* **189**, 305–14.

Armstrong, E. (1982). A look at relative brain size in mammals. *Neurosci. Lett.* **34**, 101–4.

—— and Falk, D. (eds) (1982). *Primate Brain Evolution: Methods and Concepts*. Plenum, New York and London.

Barnett, S. A. (1963). *The Rat: a Study of Behavior*. Aldine, Chicago.

Bauchot, R. (1972). Encéphalization et phylogenie. *Compt. Rend. Acad. Sci.* (*Paris*), **275**, 441–3.

—— (1978). Encephalization in vertebrates: a new mode of calculation for allometry coefficients and isoponderal indices. *Brain Behav. Evolution*, **15**, 1–18.

Bindra, D. (1976). *A Theory of Intelligent Behavior*. Wiley, New York.

Bitterman, M. E. (1975). The comparative analysis of learning. *Science*, **188**, 699–709.

Blumenberg, B. (1983). The evolution of the advanced hominid brain. *Curr. Anthropol.* **24**, 589–623.

Bok, S. T. (1959). *Histonomy of the Cerebral Cortex*. Van Nostrand-Reinhold, Princeton, N.J.

Bonin, G., von (1937). Brain weight and body weight in mammals. *J. Genet. Psychol.* **16**, 379–89.

Boule, M. and Anthony, R. (1911). L'encéphale de l'homme fossile de la Chapelle-aux-Saints. *L'Anthropologie*, **22**, 129–96.

Brodmann, K. (1913). Neue Forschungsergebnisse der Grosshirnrindenanatomie mit besonderer Berücksichtung anthropologischer Fragen. *Verhand. 85ste Versammlung Deutscher Naturforscher und Aerzte in Wien*, 200–40.

Brody, S. (1945). *Bioenergetics and Growth*. Reinhold, New York.

Bronson, R. T. (1979). Brain weight-body weight scaling in breeds of dogs and cats. *Brain Behav. Evolution*, **16**, 227–36.

Bullier, J. (1983). Les cartes du cerveau. *La Recherche*, **14**, 1202–14.

Calhoun, J. B. (1963). *The Ecology and Sociology of the Norway Rat*. United States Public Health Publication No. 1008, Bethesda, Maryland.

Clutton-Brock, T. H. and Harvey, P. H. (1980). Primates, brains and ecology. *J. Zool. Lond.* **190**, 309–23.

Crile, D. P. and Quiring, D. P. (1940). A record of the body weight and certain organ weights of 3690 animals. *Ohio J. Sci.* **40**, 219–59.

Dawkins, R. and Krebs, J. R. (1979). Arms races between and within species. *Proc. Roy. Soc. (Lond.)*, **205B**, 489–1979.

Diamond, I. T. (1979). The subdivisions of the neocortex: A proposal to revise the traditional view of sensory, motor, and association areas. *Prog. Psychobiol. Physiolog. Psychol.* **8**, 1–43.

Dubois, E. (1897). Sur le rapport du poids de l'encéphale avec la grandeur du corps chez les mammifères. *Bull. Soc. Anthropol. Paris*, **8**, 337–76.

Ebbesson, S. O. and Northcutt, R. G. (1976). Neurology of anamniotic vertebrates. In *Evolution of Brain and Behaviour in Vertebrates* (eds R. B. Masterton, M. E. Bitterman, C. B. G. Campbell, and N. Hotton), pp. 115–46. Erlbaum, Hillsdale, N.J.

Eccles, J. C. (1979). *The Human Mystery*. Springer-Verlag, New York.

Edinger, T. (1975). Paleoneurology, 1804–1966: An annotated bibliography. *Adv. Anat. Embryol. Cell Biol.* **49**, 1–258.

Eisenberg, J. (1981). *The Mammalian Radiations*. University of Chicago Press, Chicago.

Elias, H. and Schwartz, D. (1970). Cerebro-cortical surface areas, volumes, lengths of gyri and their interdependence in mammals, including man. *Zeitschrift fur Saugetierkunde*, **36**, 147–63.

Falconer, D. S. (1981). *Introduction to Quantitative Genetics* (2nd edn). Longman, London and New York.

Falk, D. (1980). A reanalysis of the South African australopithecine natural endocasts. *Am. J. Phys. Anthropol.* **53**, 525–39.

130   Harry J. Jerison

—— (1983). Cerebral cortices of East African early hominids. *Science,* **221,** 1072–4.

Friede, R. L. and Van Houten, W. H. (1962). Neuronal extension and glial supply: functional significance of glia. *Proc. Nat. Acad. Sci. (USA),* **48,** 817–21.

Garcia, J. (1975). The futility of comparative IQ research. In *Brain Mechanisms in Mental Retardation* (eds N. A. Buchwald and M. A. B. Brazier), pp. 421–42. Academic Press, New York.

Geschwind, N. and Levitzky, W. (1968). Human brain: left-right asymmetries in temporal speech region. *Science,* **161,** 186–7.

Glickman, S. E. and Sroges, R. W. (1966). Curiosity in zoo animals. *Behaviour,* **26,** 151–88.

Gould, S. J. (1966). Allometry and size in ontogeny and phylogeny. *Biol. Rev.* **41,** 587–640.

—— (1975). Allometry in primates, with emphasis on scaling and the evolution of the brain. *Contrib. Primatol.* **5,** 244–92.

Griffin, D. R. (1976). *The Question of Animal Awareness.* Rockefeller University Press, New York.

Gurche, J. A. (1982). Early primate brain evolution. In *Primate Brain Evolution: Methods and Concepts* (eds E. Armstrong and D. Falk), pp. 227–46. Plenum, New York and London.

Hahn, M. E., Jensen, C., and Dudek, B. C. (eds) (1979). *Development and Evolution of Brain Size: Behavioural Implications.* Academic Press, New York.

Harman, P. J. (1957). *Paleoneurologic, Neoneurologic, and Ontogenetic Aspects of Brain Phylogeny. James Arthur Lecture.* American Museum of Natural History, New York.

Harvey, P. H., Clutton-Brock, T. H., and Mace, G. M. (1980). Brain size and ecology in small mammals and primates. *Proc. Nat. Acad. Sci. (USA)* **77,** 4387–9.

Hawkins, A. J. and Olszewski, J. (1957). Glia/nerve cell index for cortex of the whale. *Science,* **126,** 269–78.

Hemmingsen, A. M. (1960). Energy metabolism as related to body size and respiratory surfaces and its evolution. *Rep. Steno. Mem. Hosp. Nord. Insulin Lab.* **9,** 7–110.

Herman, L. M. (1980). Cognitive characteristics of dolphins. In *Cetacean Behavior: Mechanisms and Functions* (ed. L. M. Herman), pp. 363–429. Wiley, New York.

Hofman, M. A. (1982a). Encephalization in mammals in relation to the size of the cerebral cortex. *Brain Behav. Evolution,* **20,** 84–96.

—— (1982b). A two-component theory of encephalization in mammals. *J. theoret. Biol.* **99,** 571–84.

Holloway, R. L. (1975). *The Role of Human Social Behavior in the Evolution of the Brain. James Arthur Lecture.* American Museum of Natural History, New York.

—— (1981a). Volumetric and asymmetry determinations on recent hominid endocasts: Spy I and II, Djebel Ihroud I, and the Salê *Homo erectus* specimens, with some notes on Neandertal brain size. *Am. J. Phys. Anthropol.* **55,** 385–93.

—— (1981b). Revisiting the South African Taung australopithecine endocast: The position of the lunate sulcus as determined by the stereoplotting technique. *Am. J. Phys. Anthropol.* **56,** 43–58.

—— and de la Coste-Lareymondie, M. C. (1982). Brain endocast asymmetry in pongids and hominids: Some preliminary findings on the paleontology of cerebral dominance. *Am. J. Phys. Anthropol.* **58,** 101–10.

Hopson, J. A. (1977). Relative brain size and behavior in archosaurian reptiles. *Ann. Rev. Ecol. System.* **8**, 429–48.

—— (1979). Paleoneurology. In *Biology of the Reptilia,* (eds C. Gans, R. C. Northcutt, and P. Ulinski), Volume 9, pp. 39–146. Academic Press, London and New York.

Hubel, D. H. (1979). The brain. *Scient. Am.* **241**, 44–53.

—— and Weisel, T. N. (1979). Brain mechanisms of vision. *Scient. Am.* **241**, 150–62.

Jerison, H. J. (1961). Quantitative analysis of evolution of the brain in mammals. *Science,* **133**, 1012–4.

—— (1973). *Evolution of the Brain and Intelligence.* Academic Press, New York.

—— (1976a). Paleoneurology and the evolution of mind. *Scient. Am.* **234**, 90–101.

—— (1976b). Discussion paper: Paleoneurology and the evolution of language. *Ann. N. Y. Acad. Sci.* **280**, 370–382.

—— (1977). The theory of encephalization. *Ann. N. Y. Acad. Sci.* **299**, 146–60.

—— (1979a). Brain, body, and encephalization in early primates. *J. Human Evolution,* **8**, 615–635.

—— (1979b). The evolution of diversity in brain size. In *Development and Evolution of Brain Size: Behavioral Implications* (eds M. E. Hahn, C. Jensen, and B. C. Dudek) pp. 29–57. Academic Press, New York.

—— (1982a). Allometry, brain size, cortical surface, and convolutedness. In *Primate Brain Evolution: Methods and Concepts* (eds E. Armstrong and D. Falk), pp. 77–84. Plenum, New York.

—— (1982b). The evolution of biological intelligence. In *Handbook of Human Intelligence* (ed. R. J. Sternberg), pp. 723–91. Cambridge University Press, New York and London.

—— (1982c). Problems with Piaget and pallia. *Behav. Brain Sci.* **5**, 284–7.

—— (1983). The evolution of the mammalian brain as an information processing system. In *Advances in the Study of Mammalian Behavior* (eds J. F. Eisenberg and D. G. Kleiman), pp. 113–46. Special Publication No. 7, American Society of Mammalogists.

Jerison, H. J. (1985). Animal intelligence as encephalization. *Philos. Trans. Roy. Soc. Lond. Ser. B,* **308**, 21–35.

Jones, E. G. and Powell, T. P. S. (1970). An anatomical study of converging sensory pathways with the cerebral cortex of the monkey. *Brain,* **93**, 793–820.

Kemp, T. S. (1979). The primitive cynodont *Procynosuchus:* functional anatomy of the skull and relationships. *Philos. Trans. Roy. Soc. London,* **285**, 73–122.

Kleiber, M. (1947). Body size and metabolic rate. *Physiol. Rev.* **27**, 511–41.

Lande, R. (1976). Natural selection and random genetic drift in phenotypic evolution. *Evolution,* **30**, 314–34.

—— (1979). Quantitative genetic analysis of multivariate evolution, applied to brain: body size allometry. *Evolution,* **33**, 402–16.

LeMay, M. (1976). Morphological cerebral asymmetries of modern man, fossil man, and nonhuman primate. *Ann. NY Acad. Sci.* **280**, 349–366.

—— and Culebras, A. (1972). Human brain: Morphologic differences in the hemispheres demonstrable by carotid arteriography. *New Engl. J. Med.* **287**, 168–70.

—— and Geschwind, N. (1975). Hemispheric differences in the brains of great apes. *Brain Behav. Evolution,* **11**, 48–52.

Lumsden, C. J. and Wilson, E. O. (1981). *Genes, Mind, and Culture: The Coevolutionary Process.* Harvard University Press, Cambridge, Mass.

Macphail, E. M. (1982). *Brain and Intelligence in Vertebrates.* Clarendon, Oxford.

MacLean, P. D. (1970). The triune brain, emotion, and scientific bias. In *The Neurosciences: Second Study Program* (ed. F. O. Schmitt), pp. 336–49. Rockefeller University Press, New York.

Manouvrier, L-P. (1883). Sur l'interpretation de la quantité dans l'encéphale et dans le cerveau en particulier. *Bull. Soc. Anthropol. Paris*, **3**, 137–323.

Marshack, A. (1976). Some implications of the paleolithic symbolic evidence for the origin of language. *Ann. NY Sci.* **280**, 289–311.

Martin, R. D. (1981). Relative brain size and basal metabolic rate in terrestrial vertebrates. *Nature*, **293**, 57–60.

—— (1982). Allometric approaches to the evolution of the primate nervous system. In *Primate Brain Evolution: Methods and Concepts* (eds E. Armstrong and D. Falk), pp. 39–56. Plenum, New York and London.

—— (1983). *Human Brain Evolution in Ecological Context. James Arthur Lecture on the Evolution of the Human Brain*. American Museum of Natural History.

Masterton, R. B. and Skeen, L. C. (1972). Origins of anthropoid intelligence: Prefrontal system and delayed alternation in hedgehog, tree shrew, and bush baby. *J. Comp. Physiol. Psychol.* **81**, 423–33.

McCormick, D. A. and Thompson, R. F. (1984). Cerebellum: essential involvement in the classically condition eyelid response. *Science*, **223**, 296–9.

Merzenich, M. M. and Kaas, J. H. (1980). Principles of organization of sensory-perceptual systems in mammals. *Prog. Psychobiol. Physiol. Psychol.* **9**, 1–42.

Mlikovsky, J. (1977). *Beiträg zur Evolution des Vogelgehirnes*. Dissertation, Halle/Salle, DDR (East Germany).

—— (1980). Zwei Vogelgehirne aus dem Miözan Böhmens. *Casopis pro Mineralogii a Geologii*, **25**, 409–13.

Mountcastle, V. B. (1978). An organizing principle for cerebral function: The unit module and the distributed system. In *The Mindful Brain* (eds G. M. Edelman and V. B. Mountcastle), pp. 7–50. MIT Press, Cambridge, Mass.

Nauta, W. J. H. and Feirtag, M. (1979). The organization of the brain. *Scient. Am.* **241**, 88–111.

Northcutt, R. G. (1981). Evolution of the telencephalon in nonmammals. *Ann. Rev. Neurosci.* **4**, 301–50.

Oakley, D. A. and Plotkin, H. C. (eds) (1979). *Brain Behav. and Evolution*. Methuen, London.

Parker, C. E. (1974). Behavioural diversity in ten species of non-human primates. *J. Comp. Physiol. Psychol.* **87**, 930–7.

Parker, S. T. and Gibson, K. R. (1979). A developmental model for the evolution of language and intelligence in early hominids. *Behav. Brain Sci.* **2**, 367–408.

Passingham, R. E. (1982). *The Human Primate*. Freeman, San Francisco.

Piaget, J. (1971). *Biology and Knowledge*. University of Chicago Press, Chicago.

Platel, R. (1979). Brain weight – body weight relationships. In *Biology of the Reptilia*, (eds R. G. Northcutt and P. Ulinski), Vol. 9, pp. 147–71. Academic Press, London and New York.

Plotkin, H. C. (1983). The functions of learning and cross-species comparisons. In *Animal Models of Human Behavior* (ed. G. C. L. Davey), pp. 117–34. Wiley, New York.

Quiroga, J. C. (1980). The brain of the mammal-like reptile *Proainognathus jenseni* (Therapsida, Cynodontia). A correlative paleo-neurological approach to the neocortex at the reptile-mammals transition. *J. Hirnforschung*, **21**, 299–336.

Radinsky, L. (1973). Evolution of the canid brain. *Brain Behav. Evolution*, **7**, 169–202.

—— (1977). Early primate brains: fact and fiction. *J. Human Evolution*, **6**, 79–86.

—— (1978). Evolution of brain size in carnivores and ungulates. *Am. Nat.* **112**, 815–31.

—— (1979). *The Fossil Record of Primate Brain Evolution. James Arthur Lecture.* American Museum of Natural History, New York.

—— (1981). Brain evolution in extinct South American ungulates. *Brain Behav. Evolution*, **18**, 169–87.

Rensch, B. (1959). *Evolution Above the Species Level.* Columbia University Press, New York.

Ridet, J-M. (1973). Les relations pondérales encéphalo-somatiques chez les Poissons Téléostéens. *Compt. Rend. Acad. Sci. Paris*, **276**, 1437–9.

Rockel, A. J., Hiorns, R. W., and Powell, T. P. S. (1980). The basic uniformity in structure of the neocortex. *Brain*, **103**, 221–44.

Roderick, T. H., Wimer, R. E., and Wimer, C. C. (1976). Genetic manipulation of neuroanatomical traits. In *Knowing, Thinking, and Believing* (eds L. Petrinovich and J. L. McGaugh), pp. 143–78. Plenum, New York.

Russell, D. A. (1972). Ostrich dinosaurs from the Late Cretaceous of Western Canada. *Can. J. Earth Sci.* **9**, 375–402.

Sacher, G. A. (1976). Evaluation of the entropy and information terms governing mammalian longevity. *Interdisciplinary Topics in Gerontology*, **9**, 69–82.

Sacher, G. A. and Staffeldt, E. F. (1974). Relation of gestation time to brain weight for placental mammals: Implications for the theory of vertebrate growth. *Am. Nat.* **108**, 593–615.

Scheibel, M. E. and Scheibel, A. B. (1970). Elementary processes in selected thalamic and subcortical subsystems: The structural substrates. In *The Neurosciences: Second Study Program* (ed F. O. Schmitt), pp. 443–57. Rockefeller University Press, New York.

Seligman, M. E. P. (1970). On the generality of the laws of learning. *Psychol. Rev.* **77**, 406–18.

Sharif, G. A. (1953). Cell counts in the primate cerebral cortex. *J. Comp. Neurol.* **98**, 381–400.

Simpson, G. G. (1953). *The Major Features of Evolution.* Columbia University Press, New York.

Snell, O. (1891). Die Abhäangigkeit des Hirngewichtes von dem Körpergewicht und den geistigen Fähigkeiten. *Arch. Psychiat. Nervenkr.* **23**, 436–46.

Stensiö, E. (1963). The brain and cranial nerves in fossil, lower craniate vertebrates. *Skr. Nor. Videnskaps-Akad. Oslo, Mat. Naturv. Kl.* [N.S.] No. 13, 1–120.

Szarski, H. (1980). A functional and evolutionary interpretation of brain size in vertebrates. *Evolut. Biol.* **13**, 149–74.

Szentagothai, J. (1978). The neuron network of the cerebral cortex: A functional interpretation. *Proc. Roy. Soc. (Lond.), Ser. B*, **201**, 219–48.

Thireau, M. (1975). L'allometrie pondérale encéphalo-somatique chez lez urodeles. II. Relations interspecifiques. *Bull. Mus. Nat. Hist. Naturelle*, **207**, 483–501.

Tower, D. B. (1954). Structural and functional organization of mammalian cerebral cortex: The correlation of neurone density with brain size. *J. Comp. Neurol.* **101**, 19–51.

—— and Young, O. M. (1973). The activities of butylcholinesterase and carbonic anhydrase, the rate of anaerobic glycolysis, and the question of a constant density of glial cells in cerebral cortices of mammalian species from mouse to whale. *J. Neurochem.* **20**, 269–78.

Wada, J. A., Clarke, R., and Hamm, A. (1975). Cerebral hemisphere asymmetry in humans: cortical speech zones in 100 adult and 100 infant brains. *Arch. Neurol.* **32**, 239–46.

Walker, S. (1983). *Animal Thought*. Routledge and Kegan Paul, London.

Warren, J. M. (1957). The phylogeny of maze learning. (1) Theoretical orientation. *Br. J. Anim. Behav.* **5**, 90–3.

—— (1977). A phylogenetic approach to learning and intelligence. In *Genetics, Environment and Intelligence* (ed. A. Olivero), pp. 37–56. North/Elseview, New York.

Washburn, S. L. (1978). The evolution of man. *Scient. Am.* **239**, 194–208.

Wellnhofer, P. (1974). Das fünfte Skelettexemplar von *Archaeopteryx*. *Palaeontographica,* **147**, 169–216.

Witelson, S. F. and Pallie, W. (1973). Left hemisphere specialization and language in the newborn: neuroanatomical evidence for asymmetry. *Brain,* **96**, 641–7.

Whetsone, K. N. (1983). Braincase of Mesozoic birds: I. New preparation of the 'London' *Archaeopteryx. J. Vert. Paleontol.* **2**, 439–52.

Yeni-Komshian, G. H. and Benson, D. A. (1976). Anatomical study of cerebral asymmetry in the temporal lobe of humans, chimpanzees and rhesus monkeys. *Science,* **192**, 387–9.

Models of diversity and phylogenetic reconstruction

T. S. KEMP

The question

The theory of evolution predicts that all organisms are related to one another by a unique, hierarchical genealogy; how can we hope to discover that genealogy, or more formally, how should we propose and test hypotheses of phylogenetic relationship?

The problem

There are two related problems. The first is that, on the time scale available for human observation, significant evolutionary changes do not occur, at least above the taxonomic rank of species. Therefore, the taxa available for study are effectively static, unchanging entities. The second is the possibility (most would say certainty) that many of the organisms forming parts of the real historical genealogy are unknown due to extinction and the incompleteness of the fossil record. Therefore, hypotheses of relationship between taxa can only be indirect inferences.

The solution

In order to assess the likely truth of a proposed evolutionary relationship between particular taxa, such as an ancestor-descendant relationship or a sister-group relationship, it is necessary to have ideas about the processes of evolution, such as a set of rules governing what kinds of changes are possible, how likely are convergent changes in or losses of particular characteristics, and so on. Such a set of rules constitutes a model of evolution (Cracraft 1974; Bishop 1982) and the most acceptable hypotheses of relationships will be those in greatest conformity with the model used.

On considering taxa above the species level, such models cannot be tested by direct, empirical observation of evolution in progress. Nor, of course, can they be tested by reference to the genealogies that they were used to generate in the first place. This would be tautological and therefore not acceptable within the currently accepted modes of scientific reasoning. For example, a model that states that convergent evolution is minimal can be used to generate a phylogeny from a set of taxa and that phylogeny will necessarily display minimal convergence. It would be entirely circular reasoning then to justify this particular model by observing the low incidence of convergence displayed in the phylogeny in question.

Appeal to the fossil record as a method of independently testing a model of evolution requires that the fossils actually reveal the true phylogeny in a direct, empirical manner. In practice, fossils do not form temporally continuous series (except occasionally over very brief periods). Therefore, to link fossil taxa together as 'evolving' lineages requires exactly the same kinds of assumptions about evolution as does the linking of modern taxa to hypothetical common ancestors (Kitts 1974). What fossils do potentially add to a phylogenetic analysis is a further series of taxa that need to be incorporated into the hypothesis. They may help in assessing the distribution of certain character states among taxa, but only in the same way as would extra modern taxa (Patterson 1977, 1981). The extent to which stratigraphic evidence from fossils can contribute to phylogenetic analysis is debatable. Gingerich (1979) and Harper (1976), for example, regard it as virtually essential, while others, such as Schaeffer *et al.* (1972), Patterson (1981), and Forey (1982) treat it with a suspicion bordering on total rejection. Fortey and Jefferies (1982) suggest that the confidence to be placed in stratigraphic evidence varies with the evident quality of the fossil record in question, although they do not make clear how to measure the quality of the record, without first knowing the phylogeny. For present purposes, it must be concluded that fossils are too ambiguous and incomplete to be used as independent tests of models of evolution.

Justification for certain of the models of evolution on which phylogeny reconstruction has depended has been based all too frequently either on an extrapolation of processes discovered empirically at the intraspecific level to higher taxonomic levels or upon the circular form of reasoning. In neither case can it be claimed that there is independent, empirical evidence for the validity of the model. An alternative approach recognizes the difficulties of finding independently testable models of the evolutionary process and plumps instead for a mathematical type of model, abandoning the search for evolutionary principles in favour of those of set or information theory. It is the purpose of this essay to explore the different kinds of models available and comment upon the justifications variously offered in their support.

In an attempt to dissect out and display the possible components of models of diversity, models have been classified into three categories, with subdivisions as appropriate. Brevity has demanded that each one is treated almost as a caricature, with essential features mentioned, but details and variations omitted. There is also overlap between models, in the sense that certain of them share common features and, certainly, many practising taxonomists would, if pressed, recognize components of more than one in their own particular philosophical viewpoints.

EXPLICITLY EVOLUTIONARY MODELS

(i) *Neo-Darwinian models*, based on the neo-Darwinian or synthetic theory of evolution.

(ii) *Phylogenetic cladistic models*, based on the Hennigian method of phylogenetic analysis.

(iii) *Neutralist models*, based on attempts to recognize selectively neutral characteristics, particularly macromolecular characteristics.

(iv) *Thermodynamic models*, based on non-equilibrium theories of evolution.

INFORMATION MODELS

(v) *Phenetic models*, based upon essentially mathematical ideas of how best to express information about similarity and differences of organisms.

(vi) *Pattern cladistic models*, based upon the cladistic logic, but devoid of assumptions about evolution.

STRUCTURALIST MODELS

(vii) *Structuralist models*, based upon assessment of the probability with which particular hypothetical organisms could have existed.

Evolutionary models

NEO-DARWINIAN MODELS

The neo-Darwinian or synthetic theory of evolution is founded upon the experimental and observational evidence of cellular genetics and population genetics. These lead to a hypothesis of a mechanism causing evolution, with such familiar components as random genetic variation, and natural selection of alternative alleles leading to ecological adaptation. It also embraces, if less enthusiastically sometimes, the idea of genetical sampling error of various kinds such as drift, and the genetic consequences of geographic isolation. It is thus an empirically-based theory about how genes, via their associated phenotypes, react in various kinds of ecological situations.

In order to use the neo-Darwinian theory as the basis of a model for phylogenetic reconstruction, the assumption has to be made that the neo-Darwinian processes are responsible for macroevolution, such as the origin of new higher taxa. The main components of the neo-Darwinian model of evolution applied to phylogeny reconstruction are as follows.

(i) Evolution is gradual (although not necessarily slow or continual) rather than revolutionary, matching the gradual changes in proportions of alleles observed in natural populations.

(ii) All intermediate stages are adaptive, because natural selection generating adaption is regarded as the prime mechanism of evolution.

(iii) Identical changes in separate organisms or lineages are possible, but improbable. This is because mutation of genes appears to be random with respect to DNA sequences and therefore identical changes in sequences are unlikely; these sequences control fairly directly the phenotypic characters.

(iv) Different kinds of characters vary in how likely they are to undergo convergent evolution and they can therefore be weighted in terms of the probability with which they indicate true phylogenetic relationship. Those most obviously correlated with some particular aspect of the organism's environment are most likely to undergo convergent changes, which follows from the primacy afforded to adaptation by the neo-Darwinian theory.

Simple characters are more subject to convergence than complex ones, which results from the neo-Darwinian view of the relationship between random genetic changes and phenotypic characters. Change in a complex character is assumed to depend on a large number of random mutations in many genes, and therefore to be less likely to occur independently more than once, compared to change in a simple character depending upon mutation in one or a few genes.

With components such as these, a model of evolution for use in creating phylogenetic hypotheses reads something like this: similarity arising independently is relatively uncommon, and therefore the phylogenetic hypothesis which requires the least convergence is most likely to be true. Furthermore, characters will differ in the degree to which they indicate true phylogeny. Obvious 'adaptive' characters and simple characters are more likely to be convergent. 'Non-adaptive' characters, by which is usually meant characters common to a variety of ecotypes, and structurally complex characters are less likely to be convergent. [Examples of discussions in favour of such character weighting include Cain and Harrison (1960), and Hecht and Edwards (1977), among many others]. Inferred changes from one character state to another must also be explicable as a sequence of functioning stages and therefore relationships may be based on what is judged to be the most probable such sequences (e.g., Bock 1981; Gosliner and Ghiselin 1984).

Many empirical observations have been taken as justification for the extrapolation of neo-Darwinian theory to all of evolution, but they really amount to no more than series of observations which are compatible with the extrapolation, but are not able to offer independent testing. Gould (1982) for example, has disagreed with Simpson's (1944, 1953) general thesis that the fossil record supports the neo-Darwinian view of macro-evolution. He accepts that the record is compatible with such a model, but argues that it is equally compatible with other theories of evolution. Palaeontology on its own cannot be used to discriminate between neo-Darwinian evolution and, for example, the extreme punctuationism of Schindewolf, who read the fossil record more or less literally. In the particular case of taxonomic evidence, the ability to arrange taxa into hierarchies on the basis of the distribution of their characters is certainly compatible with neo-Darwinism. However, again it is equally compatible with other conceivable evolutionary mechanisms, for example, the various versions of revolutionary change, orthogenesis, or Lamarckism. This is not to suggest that any of these other evolutionary theories are better than neo-Darwinism, but that taxonomy, at least above the species level, adds no actual corroboration to the neo-Darwinian theory, compared to other possible theories. It follows, therefore, that a model of evolution based on neo-Darwinism can only be justified by deductive arguments to the effect that the mechanism of evolution proposed by the neo-Darwinian theory, given enough time, will indeed produce the higher taxonomic patterns. Herein lies a current evolutionary debate: the extent to which empirically observed events at the intraspecific level really can be extrapolated to account for the higher level evolutionary events of macroevolution.

Several population geneticists (e.g., Charlesworth *et al.* 1981; Maynard Smith 1983; see Levinton 1983 for a recent review) have defended extrapolation, arguing that nothing has been adequately demonstrated about macroevolution that refutes the neo-Darwinian theory and therefore there is no need to abandon it when explaining macroevolution. Indeed, they would no doubt argue that it is bad scientific practice to drop a theory with at least some empirical backing, until it ceases to account for observations. Conversely, several other recent authors have explicitly questioned whether extrapolation can account for patterns of macro-evolution. Their objections can be traced originally to the apparently punctuated pattern of the fossil record, new species appearing instantaneously, and thus suggesting that the gradualism of evolution predicted by neo-Darwinism was refuted (Eldredge and Gould 1972; Gould and Eldredge 1977; Gould 1980; Stanley 1975, 1979; see Hoffman 1982 for a critique of the palaeontological evidence).

Since then, a number of possible evolutionary mechanisms have been pointed out that are not part of the strict neo-Darwinian theory, but for which there is at least some experimental evidence. Important evolutionary events could be a result of random chromosomal mutations, causing instantaneous, non-adaptive speciation (e.g., Bush *et al.* 1977) or large phenetic change. Mutations in regulator genes, such as homeotic mutations, which can cause large, but functionally co-ordinated phenotypic change have evoked a lot of interest as a possible evolutionary mechanism (e.g. Raff and Kaufman 1983). The evolutionary modifications that some particular characteristic can undergo may be constrained to a very limited number, due to the nature of developmental processes (e.g., Rachootin and Thomson 1981). If this were so, then change would not be completely random, and convergent change could be at least common, if not virtually inevitable. Also along these lines, it appears quite possible that different mutations in different genes may trigger the same modifications to development, leading to the same phenotypic changes.

All these alternative possible mechanisms of evolution bear upon the idea of *a priori* character weighting, which is so central in practice to the neo-Darwinian model. Convergent evolution of a complex character may be common because the change depends upon a simple genetic trigger. A simple character may alter very rarely because of excessive developmental constraints on its possible evolution. The point is that not enough is known about the genetic basis of phenotypic characters to predict the probability of particular changes beforehand (Eldredge and Cracraft 1980).

At an ecological level, Stanley (1981) has produced interesting, if rather anecdotal evidence for the possibility of relatively inadaptive variants remaining viable because of fortuitous isolation. The possibility therefore exists that poorly adapted stages may occur during the evolution of some particular group. Cracraft (1981), in particular, has criticized the idea that the probability of a relationship between two taxa can be judged by whether a series of functionally well adapted intermediate forms can be imagined. There is no certainty that such a series need always have existed.

The idea that a species, or even a higher monophyletic group of species, can behave as an individual has been developed (Ghiselin 1974; Gould 1980; Vrba and Eldredge 1984; Vrba 1984). Such higher order individuals may have their own emergent properties, such as character sizes, origination rates, and extinction rates. More dubiously, they may posses group characters capable of endowing the group as a whole with relative survival values, so that a natural selection process between monophyletic groups may occur (Stanley 1975, 1979; see Ayala 1983 and Maynard Smith 1983 for critiques). The consequences for evolutionary theory of what has come to be termed decoupling of macroevolution from microevolution (Gould 1980), is that macroevolution cannot be explained by the neo-Darwinian processes, but that such phenomena as the origin of new taxa, evolutionary trends, etc., require explanation by other, higher order processes such as differential species survival.

The point at issue here is not which of these various possible mechanisms of evolution are correct, but that a number of alternative mechansims to the ones assumed by the neo-Darwinian theory have been proposed, have at least some empirical basis, and may therefore be correct. If a neo-Darwinian model of evolution is used as the basis for phylogenetic reconstruction, then the phylogenies generated may be wrong, to the extent that the model is unrealistic.

The same comments will, of course, be true of any specific evolutionary model (for example, a model based on the idea of random speciation, or on a decoupled theory of evolution). Other authors have suggested alternative evolutionary models in which the criteria for characters weighting differ. Løvtrup (1975, 1977) suggested a model in which morphological characters are given very little weight compared to biochemical characters. He defended this on the grounds that small developmental changes could produce large morphological ones, so morphology would not be a reliable indicator of relationship. Needless to say, this model is as incapable of direct empirical testing as any other evolutionary model, pending a great deal more understanding of the potential of particular kinds of characters for particular kinds of evolutionary change.

PHYLOGENETIC CLADISTIC MODELS

The cladistic method of phylogenetic analysis was introduced by Hennig (1966) as an attempt to improve the objectivity, and therefore testability, of phylogenetic hypotheses. As originally perceived, it appeared to depend upon three assumptions about the process of evolution (Cracraft 1974). The first is that macroevolution occurs by dichotomous splitting of a single species into two, at which time (and only then) the ancestral species becomes extinct. The second is that the members of the two descendant or sister groups will possess respective evolutionary novelties, i.e., derived or apomorphic characters. Hennig stressed that the only empirical evidence for genealogical relationship is the common possession of such derived characters, or synapomorphy, and that there are criteria for recognizing synapomorphy with an acceptable degree of confidence. The third is that convergence is relatively rare and therefore the best hypothesis of phylogeny will be that implying the least number of such events.

The extent to which these assumptions can be (and to some extent probably always have been; Hull 1979) regarded as logical procedures for ordering taxa, rather than as strict beliefs about how evolution occurs, will be discussed later. For the moment, it may be stated that if a cladogram is to be a true representation of phylogeny, then the assumptions must necessarily be correct.

[Much of the phylogenetic cladistic programme is concerned with the correct method of expressing a reconstructed phylogeny as a formal classification, and this is also the area about which there is most dispute (e.g., Hull 1970; Mayr 1981). However, such matters are not the concern of this paper, as we are only concerned with the problem of phylogenetic hypotheses *per se*.]

There has been a great deal of argument about the cladistic assumptions as a basis for a model of evolution. Evolutionary biologists, particularly those concerned with systematics at and below the level of the species, have been greatly exercised by the possibility of a single species undergoing a multiple split, by a single species remaining unchanged after giving rise to a descendant species, by the possibility of such a species producing a temporal succession of new species and so on. If these sorts of mechanisms are possible at lower taxonomic levels, as presumably they are, then there is no reason to expect that the higher taxa, which ultimately derive from such speciation events, should fall into neat dichotomous hierarchies.

The cladistic view has also been accused of treating taxa as Aristotelian-like classes of entities, membership of which is determined by the possession of defining characters. This is at variance with an evolutionary interpretation of a taxon, whereby it is an individual possessing a history of change from origin to extinction, and therefore no particular character needs to occur in all its members (e.g., Ghiselin 1974; Hull 1976; Beatty 1982). The 'defining' characters may be yet to evolve, or have been subsequently lost in a particular member of the taxon, without necessarily excluding that individual from the taxon. An extension of the same philosophical argument about the unreality of the cladistic model of evolution concerns its attitude to ancestry. Taxa can only be recognized by their derived characters. By definition, an ancestral taxon does not possess derived characters (that is, derived relative to its descendants), and therefore a real ancestral taxon cannot be recognized. Yet to all evolutionists, ancestry must be a real, not to say fundamental concept.

However, leaving aside these specific criticisms of the evolutionary asumptions of phylogenetic cladism, the main difficulty is that of all evolutionary models. The assumptions it makes about the processes of evolution, at least above the species level, are empirically untestable. The true phylogeny may involve multiple splitting of ancestral taxa, large amounts of convergence, and secondary loss of characters, but this would never be expressed as a result of using the model. Many cladists acccpt these criticisms, and have moved, to varying degrees, towards the transformed or pattern cladistic postition discussed in a later section.

NEUTRALIST MODELS

The Holy Grail of phylogenetic reconstruction has always been the hope of discovering characters which are unaffected by natural selection, and which also have a small, but positive probability of random change. They would be expected to be virtually free of convergence, and would faithfully track phylogeny. From time to time the belief has been expressed that complex morphological characters associated with major body plans, and which remain constant in a wide variety of organisms in different habitats, are just such characters. However, such assertions can be criticized on the same grounds as criticism of character weighting in general, namely that there is no independent empirical test that these characters are not subject to convergence. For example, it is possible that a particular character could only undergo a very small range of possible changes, perhaps only one, due to developmental constraints. Convergent change would thereby be highly likely. Furthermore, the hypothesis that a particular character is selectively neutral or non-adaptive is extraordinarily difficult to confirm in practice. Even a character with no currently known function may be adaptively important in some subtle physiological way not discovered, or may possibly play a role during epigenesis, even though non-functional in the adult. Notwithstanding Gould and Lewontin's (1979) criticism of over-enthusiastic searching for the adaptive significance of each and every attribute of an organism, an assumption of the non-adaptivness of a particular character is certainly no more acceptable (e.g., Rudwick 1964).

   The description of comparative macromolecular sequences during the last decade or so has, however, re-opened the whole question of adaptively neutral characters. What has become known as the neutral theory of evolution asserts that most mutations in DNA, as represented directly by different nucleotide or indirectly by amino acid sequences in different organisms, are selectively more or less neutral, and accumulate by random processes of genetic drift (King and Jukes 1969; see Kimura 1983 for a review). The empirical basis for this theory is the large amount of variation seen in the amino acid sequences, and hence presumably in the alleles within living populations of virtually all organisms studied (e.g., Ayala 1982). The theory can form a model of evolution for use in phylogenetic reconstruction, making use of sequence data characters. Despite its theoretical promise, the sequences have proved less unambiguous than hoped for (e.g., Goodman 1982). Referring to a single protein, phylogenetic hypotheses cannot be discovered which are anywhere near free of implied convergent and/or reverse mutations. Taking several proteins separately leads to a series of different best hypotheses of relationship for the same set of taxa. Therefore, notwithstanding the claims of the theory, convergence and reversal affect protein evolution as they do morphological evolution. As soon as this is recognized, then the same problem of independent justification of this model arises as applies to all models of evolution. If some mutations occur convergently, either as an adaptive response, or possibly because certain mutations are more probable than others for chemical reasons, then it is difficult to see how any mutation can confidently be described as neutral and random in the absence of prior knowledge of the phylogenetic relationships. Also, rightly or wrongly, the very empirical basis of the neutral theory has been attacked by several population geneticists, for example McDonald (1983), who argue that the large genetic variation in real populations can be accounted for in adaptive terms.

THERMODYNAMIC MODELS

From Lamarck's perfecting principle through concepts such as Osborne's creative aristogenesis and Schindewolf's autonomous laws, there has been a heterodox tradition to the effect that evolution is directed by higher order laws than natural selection (for a review see Rensch 1959). If ever such laws were indeed discovered, then a phylogenetic hypothesis could be tested in principle by the extent to which it conformed to them.

The most recent version of the tradition views evolution as an irreversible process subject to the second law of thermodynamics, and which is therefore constrained to go in a direction that increases entropy. Wicken (1981, 1984) argues that the generation of increasing complexity of organisms during evolution corresponds to a decrease in potential energy and an increase in information entropy. It is what a physicist would describe as a dissipative process, with an irreversible, unidirectional flow over time. It is claimed that many aspects of prebiotic evolution can be understood in these terms and Wicken sees no reason to doubt that the same physical laws continue during biotic evolution.

Wiley and Brooks (1982; Brooks and Wiley 1984) have also developed the idea of evolution as a non-equilibrium, dissipative process. To them, there is an increase in informative entropy as variation increases, and also in what they term cohesion entropy, as the uniform breeding system of a species breaks down during speciation.

At the present time, the whole question of whether evolution can be treated as a process subject to the laws underlying non-equilibrium dissipative processes in the physico-chemical world is debatable (Wicken 1983; Bookstein 1983; Løvtrup 1983; O'Grady 1984). Despite the independent contribution that these studies could theoretically make to phylogenetic reconstruction (O'Grady 1984), it must be doubtful whether anything of practical benefit will emerge.

Information models

Recognition of the frailty of the explicitly evolutionary models available for converting empirically derived taxonomic patterns into phylogenetic hypotheses has led to a number of attempts to exclude as much assumption about evolutionary processes as possible. Taxa, and their variously distributed characters, are taken as a form of biological information and treated simply as that, applying the ideas of and methods for handling, storing, and retrieving generally. Even the basic assumption that evolution is the cause of diversity may be rejected as part of the methodology (although this should not be confused with the personal philosophical beliefs of the exponents of these methods, of course!).

As will be shown, however, information methods are actually no freer of assumptions than evolutionary methods, although the assumptions are of a different kind.

Information models

PHENETIC MODELS

The phenetic approach to the analysis of the diversity of organisms explicitly excludes all prior assumptions about the evolutionary process. This includes abandoning *a priori* character weighting, with its inbuilt assumption that the different evolutionary properties of different kinds of characters can be recognized, as well as the more obvious assumptions such as rarity of convergence. The main principle invoked is that the relevant information in a classification is simply the distribution of similarities and differences of attributes amongst the organisms in question, and therefore the most informative classification will be the one that incorporates the largest number of statements of similarity (and, conversely, difference). Groups or taxa recognized in this way are said to be natural in the sense of Gilmour (1961; see Farris 1983; Panchen 1982), that is, groups about which the greatest number of general statements may be made.

To achieve this end, a statistic of degree of similarity between each pair of taxa involved is estimated, by using a large number of unweighted characters and deriving an average 'morphological' distance between them. The phenetic classification is generated from these statistics by clustering more similar taxa together, less similar taxa apart. There is actually a wide variety of detailed methods available, differing in the nature of the algorithms applied both to convert the raw data into a phenogram and to form clusters.

A secondary principle almost always adopted is that the most informative kind of taxonomic pattern is a hierarchical arrangement of taxa, that is, the taxa arranged in non-overlapping, internested sets. Sneath and Sokal (1973) admit that the reason for seeking a hierarchical arrangement may have more to do with human psychology than with whether such a pattern is in any way the 'real' pattern.

Criticism of phenetic methodology because it does not include even 'reasonable' evolutionary assumptions (e.g., Hull 1970) is not relevant at this point. Of course, a phenogram may be read as a phylogenetic hypothesis with the addition of an appropriate set of evolutionary assumptions (Sneath and Sokal 1973), but the necessary assumptions are as open to criticism as those already discussed. There are, however, several difficulties to phenetics even within the context of its professed aim of recognizing a totally empirical taxonomic pattern that is free of assumptions about how the pattern arose. First of all, there is the serious problem of how to actually achieve an objective measure of overall phenotypic similarity. It is not at all clear, for example, exactly what constitutes a unit phenotypic characters (Crowson 1970), or how to deal with an array of taxa which have widely different numbers of characters available.

A more fundamental difficulty discussed by Farris (1979) arises from the use of large numbers of equally weighted characters. The method allows a character that is already shared by two taxa to contribute to the overall degree of similarity between them, even though that particular character may also occur in other taxa. A phenetic group is therefore defined in part by possession of attributes not unique to that group. Since these particular attributes could, with equal logic, be part of the definitions of different groupings, they are irrelevant and must be regarded only as 'noise' rather than 'signal'. If there is a high percentage of characters of this nature made use of, then the noise level may very well drown the signal, and the resulting classification be virtually random. Pheneticists such as McNeill (1982) may argue in effect that if there is a very large number of 'noisy' characters, then the signal characters will stand out against a random background of irrelevancy. This would be true if there were grounds for believing that the 'noisy' characters will be distributed randomly amongst all the taxa, but there is no self-evident reason why this should be true. In any case, the pattern that would emerge against the background would actually approximate to the pattern cladist kind of pattern discussed shortly. It further follows from this argument that the level of predictability, or the information content, of a phenetic classification is limited. No character is either necessary or sufficient for indicating membership of any particular group, and therefore no member of a given taxon can be stated with certainty to possess any particular character simply by virtue of its membership. Such is the polythetic nature of Gilmour-natural groups as conceived by phenetics. Whether this is a critical difficulty or not of the phenetic model depends on one's view, but rejection of phenetic models on these grounds leads to an alternative kind of information model.

PATTERN CLADISTIC MODELS

The original phylogenetic cladistic model of evolution has already been seen to be no better, and in many eyes a great deal worse, than a neo-Darwinian model as a basis for phylogenetic reconstruction. It possessed such dubious assumptions about the evolutionary process as dichotomous branching, treatment of taxa as Aristotelian classes with defining character-istics rather than as individuals with their own histories, and it had, to say the least, a somewhat ambiguous attitude to the reality of ancestry.

However, cladism has been saved from the fate of being labelled merely another empirically untestable model of evolution by a remarkable observation. Platnick (1979; see also Hull 1979; Nelson and Platnick 1981) pointed out that the logic and procedure of phylogenetic cladism correspond not only to one particular model of evolution, but also to a certain set of precepts appropriate to a high resolution method of general classification of a diversity of entities. Re-expressed in non-evolutionary terms, the cladist approach has become termed transformed or better pattern cladism (Beatty 1982).

The central point about pattern cladism is rejection of the idea that the natural (i.e., 'real') groups of organisms are the Gilmour-natural groups of phenetics, because of the ambiguity of such polythetic groups. Instead pattern cladists prefer to accept as natural only those groups whose members all possess features otherwise unique to the group, that is to say, characteristics both necessary and also sufficient for membership. Farris (1979) expresses it as the use of special similarity rather than raw similarity for defining groups. The result is a set of completely non-overlapping groups or taxa. It is an empirical observation that characters vary in the degree of generality with which they occur amongst organisms. Relatively widely distributed characters (such as vertebrae) define unique groups with large numbers of members (Vertebrata). Less widely distributed characters (such as hair or feathers) define more restricted groups (Mammalia and Aves) within the larger groups. Therefore, an internested or hierarchical arrangement of group, and subgroups emerges. With a hierarchical pattern such as this, the greatest number of different sets of characters, defining the greatest total number of included groups, is found in a dichotomous hierarchy, where each group is divided into two subgroups, and so on (Platnick 1979). Such a pattern can be described as completely resolved, or as containing the greatest amount of relevant taxonomic information.

In an ideal world, all the characters would define the same set of Aristotelian groups, each individual character being taken at its appropriate level in the hierarchy. In other words, there would be total congrunece of distribution of all the characters. As this is manifestly not the case with organisms, some characters indicating one set of groups, other characters indicating a different set, there is the problem of deciding which particular set of groups is the best. Since total congruence is the ideal, the pattern regarded as most likely to be the natural pattern is the one showing the highest degree of congruence, which is the pattern supported by the largest number of characters.

The three major procedures of pattern cladism correspond in logic, if not in terminology, to the main assumptions of phylogenetic cladism. Thus, definition of natural groups by unique characters corresponds to the recognition of monophyletic groups by possession of uniquely derived characters; the dichotomous hierarchical arrangement corresponds to the dichotomous splitting of lineages; and maximum congruence corresponds (at least at first sight; see below and Felsenstein 1983; Farris 1983) to rarity of convergence.

In the few years since its introduction, pattern cladism has attracted a lot of adverse comment, as did phenetics before it. Most of the criticism is not of the method in its own terms, but to the effect that the terms are wrong and taxonomists should include evolutionary assumptions in their reasoning. Beatty (1982) and Ghiselin (1984), for example, complain of taxa being treated as Aristotelian classes with defining properties, rather than as individuals with histories of change. However, to regard an observed taxon as an individual requires the introduction of an evolutionary assumption, namely, that organisms and taxa change with time. Patterson (1982) suggested in reply to Beatty, that the only way to recognize a taxon in practice is by the presence of character features, whether one thinks of it philosophically as a class or as an individual. Charig's (1982) attack is similarly an argument that evolutionary assumptions should be used in taxonomy, not a discussion of whether pattern cladism achieves its self-proclaimed purpose of excluding empirically untestable assumptions.

There are, however, several areas in which pattern cladism may be challenged in its own terms. The first is the question of what assumptions are used to recognize the rather less than self-evident concept of a unit character, a problem common to all information models of diversity.

A more fundamental question is why the natural order of nature should be assumed to be a hierarchy of non-overlapping groups (Bremer and Wanntorp 1979). Non-congruence, or homoplasy as it is usually termed, is not adequately explained by pattern cladism, except somewhat lamely as a result of mistakes in the analysis. It is presumably hoped that the errors will be discovered and removed. However, a really strict empiricist would argue that homoplasy is as real, or natural, a phenomenon as homology, since it is there to be seen. In the absence of evolutionary assumptions, it could then be argued that the 'natural' pattern is not hierarchical but reticular, consisting of groups arranged as overlapping Venn diagrams in multidimensional space. A given taxon could be a member of one group by virtue of possession of certain characteristics, and at the same time belong in other groups by virtue of possession of other characters not congruent with the first set. In theory, the most fully resolved classification would then consist of a definable group for each and every possible combination of organisms, a multi-dimensional network of groups. Absolutely every character would be incorporated, and there would cease to be anomolies such as non-congruent characters. Naturally, a reticular system would not be very convenient for use as a biological classification, which is no doubt why it has never been adopted, even in pre-evolutionary times [see Nelson and Platnick (1981) for a history of rudimentary attempts at this approach]. It may be noted, nevertheless, that there is a well-established human use for reticular classifications as the basis of information storage and retrieval. They are called cross-reference systems and are extraordinarily useful for books or museum specimens. Certainly they are empirical, with no obvious assumptions about the processes generating the diversity in question. In short, one should ask why Aristotelian groups should be assumed *a priori* by pattern cladism, when immediate experience of the diversity of organisms indicates that overlapping, polythetic groups are what one often sees.

The pattern cladist assumption that maximum congruence generates the closest approximation to the natural pattern has been criticized by Farris (1983) on the grounds that it is not necessarily the same as minimum homoplasy. Those characters which are simply excluded because they do not fit the pattern in congruence with a majority of other characters may actually imply a large number of separate instances of homplasy (i.e., 'mistakes') in the cladogram as a whole. A reduced total number of incidences of homoplasy may be discovered by assuming that characters other than those forming the most congruent set are the correct homologies defining the groups. Whether maximum congruence or minimum homoplasy should be adopted seems to depend on which particular *a priori* assumptions about the natural pattern are made.

Pattern cladist models cannot be described as free of theory, hypothesis, or assumption any more than evolutionary models. Assumptions about the processes causing diversity may be excluded, but it is still necessary to decide beforehand what form the natural pattern should take. What constitutes a unit character? Overlapping or non-overlapping groups? Hierarchy or reticulum? Congruence or minimum homoplasy? As much as any evolutionary taxonomist, the kind of 'natural' pattern that an information taxonomist sets out to discover is assuredly the kind that he will find.

Structuralist models

Biological structuralism is a term applied in recent years to the biological viewpoint corresponding more or less to idealistic morphology (see Mayr 1982 and Reif 1983 for historical discussions; Riedl 1978; Goodwin 1984; Webster 1984 for modern views). It is therefore a modernized version of a persistent, if frequently reviled (Mayr 1982), strand of thinking with roots dating well back to before 1859. Structuralism is concerned with discovering the laws that dictate the form of organisms, and the ontogenetic processes which generate it. It excludes consideration of any processes, evolutionary or otherwise, by which form has come about historically, on the grounds that while form may be studied empirically, the historical generation of form cannot. Indeed, structuralism relates to the characters of organisms in an analogous manner to the way in which an information model such as pattern cladism relates to taxa. A structuralist is interested in many kinds of problems. How are the different characters or attributes of an individual organism organized relative to one another (e.g., Riedl 1978)? What forms can and cannot the morphogenetic mechanisms of an organism produce (e.g., Goodwin *et al.* 1983)? What are the design principles underlying organism structure (e.g., D'Arcy Thompson 1942 for a classic study)? Overall, one of the desired outcomes of the structuralist research programme is a knowledge of what phenotypic structure is possible and what is not.

As applied to the problem of understanding the diversity of organisms, a structuralist approach is really a method that could be used with either evolutionary models or information models of the kinds already discussed. In both cases, it offers a theoretical possibility of providing tests of hypotheses of pattern. If laws of structure were to be established, they would dictate which sets of characters or attributes could co-exist in an organism and which could not. In the terms of an evolutionary model, the hypothetical ancestors necessarily implied by a phylogenetic hypothesis could be inspected. If any of them consisted of sets of characters that contradicted the laws, then the hypothesis which generated them would be refuted. In the case of an information model, a hypothesis of pattern implies that certain sets of characters have the same level of generality (i.e., co-exist in the same group). For example, in the case of a hierarchical pattern cladogram, the association of two taxa as sister groups implies a certain set of characters represented by the node between them. This set will be a combination of the characters unique to the two taxa in question, along with the more general character states of those characters in which the two differ. If the laws of structure forbid the co-existence of that particular set of characters, then the hypothesis is refuted.

Theoretically, structuralism offers a possible test of whether a hypothetical phylogeny or pattern could be true, which is independent both of unverifiable assumptions about evolutionary processes and of appeal to statistical methods of unknown biological significance. If it happens that all, or even most, imaginable phenotypes could actually exist, then it has to be admitted that any hypothesis could be true and therefore that phylogeny (or 'true' pattern) becomes, in principle, unknowable (which is not necessarily to say that some hypotheses cannot be judged more likely than others). If it is true, however, that much of imaginable 'morphospace' is prescribed, then the possibility of refuting certain hypotheses of phylogeny (or pattern) exists.

In practice, of course, structuralist laws have been long promised, but precious little has yet been delivered, and models of diversity based on them remain a matter of speculative possibilities. Nevertheless, there are several conceivable areas from which laws may emerge. At the macro-molecular level, certain sequences of nucleotides or amino acids may prove to be chemically impossible or, as least, highly unlikely. At the developmental level, the concept of constraints suggests that there may be no possible epigenetic mechanism for producing certain forms. At the level of the whole organism, there are several ideas around to the effect that structure is constrained by laws, so that certain imaginable morphologies cannot exist (e.g., Riedl 1978). There is also the very real phenomenon of functional integration of organisms, whereby all the parts and processes of a single organism are interrelated in networks of causes and effects. Therefore, a hypothesis could be judged wrong if it implied a set of characters shown to be functionally incongruent.

Discussion

Methods for discovering the biologically significant patterns of diversity of organisms have been seen to fall into two principal categories.

Patterns (called phylogenies) based on evolutionary models. The model adopted depends on which particular microevolutionary processes are chosen to represent, by extrapolation, the cause of macroevolution. There is a wide range of feasible processes suggested which are not always compatible with one another, but none of them has been corroborated beyond the species level and therefore the basis for choosing one, but not another is unclear.

Patterns (called taxonomies) based on non-evolutionary models. The model adopted depends on selecting an uncorroborated hypothesis about which particular kind of pattern is 'natural'. Should groups be monothetic or polythetic, arranged hierarchically or reticularly, based on maximum congruence or minimum homoplasy?

> (Patterns based on structuralist models offer a method that must nevertheless be used in conjunction with either an evolutionary or an information model, and are not therefore regarded as a third, independent kind of model. In any case, they depend on the uncorroborated hypothesis that not all morphologies are equally possible.)

It is clear that the major controversy within the taxonomic community concerns the question of which is the proper position to occupy along a scientific philosophical spectrum, a very ancient debate indeed (e.g., Harré 1972). At one extreme of the spectrum lies the phenomenalist who accepts as real only his immediate sensory input from the natural world. Ghiselin (1984) has suggested that pattern cladists occupy this position, a view that is hard to maintain in the light of their active search for real patterns underlying the immediate phenomena. Indeed, a true phenomelalist would hardly accept such hypothetical constructs as homologous characters or taxa, let alone higher order, hierarchical patterns (M. Greene, pers. comm.). In fact, non-evolutionary taxonomists may be designated naive realists, accepting the reality of taxonomic patterns that can be hypothesized with confidence, i.e., with a minimum of uncorroborated assumptions. Evolutionary taxonomists lie still further along the spectrum, for they are prepared to accept the reality of causes of the patterns. Their hypotheses of cause, or explanations, demand further assumptions of increasingly speculative rather than directly testable nature. Nelson and Platnick (1984), for example, described (and dismissed) evolutionary taxonomy thus:

> After 1859, such [ancestral] taxa and areas were soon 'discovered' and the literature is now replete with them. We argue that such taxa and areas were discovered for no other reason than because the theory, as conceived or misconceived by Darwinians, demanded them. We argue that the taxa and areas are not phenomena, but rather artifacts. We argue finally that the theory that demands them is false and that the falsifying evidence presently available is decisive. . .

On the other hand, Charig (1982, p. 436 expressed a somewhat similar view, but drew the opposite conclusion.

> I recognize that we have no absolute proof of the theory of evolution, by direct evidence of the senses; all the available evidence is merely circumstantial. However, there is no scientifically acceptable evidence against evolution and no other theory fits the known facts so well. I therefore accept it as a working hypothesis of immense heuristic value.

The question of which is the better philosophical stance to take in the context of understanding and expressing the diversity of organisms is difficult to decide. Indeed, in everyday practical terms it probably does not matter very much. An evolutionary biologist, who sees the inherent difficulties of justifying *a priori* character weighting and of recognizing ancestors, and an informational taxonomist, who accepts that the natural pattern is hierarchical and has come to terms with the problem of non-congruence instead of effectively ignoring it or hoping it will average out, are likely to produce virtually identical patterns of taxa from the same character distributions. The main difference between them will be the terminology they use to express the pattern: monophyletic groups versus natural groups and so on.

Most of contemporary biology expresses the results of its investigations in explanatory terms, to the effect that the patterns, correlations, causes and effects, and so on discovered, are the result of real underlying mechanisms. Theories and hypotheses are couched in the terms of these assumed processes. There seems no sensible reason to abandon such a philosophy in one particular branch of biology such as systematics. Equally, however, it would be wrong to fall into the habit of regarding uncorroborated or even weakly corroborated assumptions about evolutionary processes as received certainties. While the general idea of evolution is the most powerful mechanistic explanation of diversity of organisms, indeed there are no very sensible alternatives, any particular hypothetical mechanism of evolution is considerably less powerful because several plausible alternative mechanisms do exist.

THE CONTEMPORARY APPROACH

In this light, the most important thrust of current theoretical taxonomy consists of exploring the properties of models which combine a limited and precisely stated set of assumptions about evolution, with general statistical methods aimed at discovering the phylogenetic hypothesis which best explains the character distributions. The general procedure is formally termed likelihood, in which the likelihood of a hypothesis is a measure of the amount of evidence supporting it (e.g., Edwards 1972). It is not to be confused with a measure of the probability of the hypothesis being true, because without first knowing the truth it is impossible to make such an estimate. Rather, it is a measure of the extent to which the hypothesis explains the available observations, without prejudice as to whether those observations are correct or representative. In the case of phylogenetic reconstruction, the likelihood of a particular phylogenetic hypothesis of relationships depends on the evolutionary assumptions made about such things as probabilities of change of characteristics, as well as the simple distribution of characters.

Sober (1983) analysed the properties of a very simple model of the evolutionary process, based upon the following assumptions: (i) character states can be recognized as either plesiomorphic (ancestral) or apomorphic (derived); (ii) the probability of transition from plesiomorphic to apomorphic states always has a probability between 0 and 1 (that is, any proposed character change is neither certain nor impossible); (iii) evolutionary change in one part of the phylogeny is independent of change in any other part; (iv) the probability of a character changing from plesiomorphic to apomorphic states is lower than the probability of an apomorphic characteristic remaining unchanged.

Under the terms of this model, the maximum likelihood phylogeny can be shown to be the one that has the most parsimonious distribution of apparent apomorphies, i.e., the one that incorporates the least homoplasy (convergence/loss). Contrary to intuitive reaction perhaps, this does not mean that evolution is assumed actually to have been parsimonious, generating minimum homoplasy, only that while the hypothesis may be wrong because evolution was not, in fact, parsimonious, nevertheless there is no way of recognizing a better hypothesis to explain the distribution of the particular characters available. Farris (1983) has presented the clearest argument for this interpretation of the method. Suppose ten apparently apomorphic characters support one grouping and one apomorphic character supports another. Put simply, only one apparent apomorphy needs to be true to establish the true relationships. Therefore, there are ten chances of the first hypothesis being correct, but only one chance of the second being correct. In the absence of any special information about the particular characters, such as their respective rates of change, the first hypothesis has a higher likelihood of explaining the character distribution and should therefore be preferred.

A model such as this one makes about as few assumptions of evolution as possible. Other authors, notably Felsenstein (1978, 1983; see Friday 1982) have been interested in the statistical analysis of character distributions using different evolutionary assumptions, when parsimony may not produce the maximum likelihood phylogeny. Felsenstein's best known example is a case where it is assumed that the probability of change differs in different parts of the phylogeny. Here the probability of unrelated lineages evolving apparent synapomorphies by convergence may be higher than the probability of related lineages possessing apomorphies. As more and more apparent synapomorphies are discovered, the unrelated forms will accumulate increasing numbers of shared convergences and therefore appear increasingly to form a monophyletic group. The true monphyletic group will be increasingly obscured. Critics of Felsenstein (e.g., Farris 1983) argue that such detailed models as this are unrealistic and therefore should not be adopted. In our present state of ignorance, parsimony procedures, which involve less assumptions, remain preferable. It is perfectly true that this kind of approach to phylogenetic reconstruction still appears to depend on essentially untestable assumptions about the evolutionary process, however minimal. The advance they make lies in their precision, both in terms of formulating exactly which assumptions have been made and in explicating the nature of the logical processes applied. If and when we do begin to grow more confident about the mechanisms of macroevolution, these kinds of models will be needed.

CONCLUSION

It is an extraordinary feature of systematic biology that the arguments between apologists for the various schools are often conducted at a most passionate level. Perhaps this is partly because 'belief' or 'non-belief' in evolution has spilled out of the scientific framework into politics and religion. More importantly perhaps, it has to do with the way that the whole of biology has become couched in evolutionary terminology. This includes the areas of biology less obviously concerned with theoretical evolutionary matters and involving biologists who do not need to consider evolutionary theory themselves. Expressions like 'adaptation', 'related organisms', 'the primitive condition', etc., are included in the jargon of areas as far apart as molecular biology, physiology, and ecology, whether they are justified in each particular instance or not. The consequence has been a loss of distinction in much of biology between concepts, underpinned by strongly corroborated hypotheses, concepts only weakly defensible, and concepts which are no more than tentative speculations.

On the one hand, evolutionary taxonomists mistakenly see the threat posed by information taxonomists to their terminology as a threat to rational biology as a whole. On the other hand, information taxonomists fear for a science that is embedded too unconciously in universal assumptions, many at least of dubious value.

I have tried in this all too brief essay to show that all attempts to understand the diversity of organisms rely upon empirically untestable assumptions either about evolution or about natural patterns. There is nothing wrong with making assumptions nor seeking to justify them of course. It is the very stuff of science. What is unforgiveable is to forget that they are assumptions and behave as if they are known certainties when they are no such things.

Never has the need for careful understanding and explanation of the basis of models of diversity been greater, for new information about diversity, and its underlying genetic and epigenetic causes is flooding in, particularly from the areas of biology concerned with macromolecules. There are ever more sequences of DNA and proteins being published. Unexpected intragenomic events such as transposable elements, RNA editing, and silent DNA are being described and related to evolutionary mechanisms (Dover and Flavell 1982). The relationship between the genome and developmental mechanisms producing the phenotype is under active scrutiny (Raff and Kaufman 1983). Our models need to take all this into account if we are to hope to home in on the true meaning of taxonomic pattern.

Acknowledgements

I am very grateful to various colleagues for reading and extensively discussing earlier drafts of this paper, particularly Mike Benton, Gillian King, Dave Norman, Malcolm Scoble, and Keith Thomson. Marjorie

Grene was especially helpful on the philosophical issues and the editors, Richard Dawkins and Mark Ridley, have been kind enough to publish the paper despite my inability to persuade them of the truth of every word!

References

Ayala, F. J. (1982). The genetic structure of species. In *Perspectives on Evolution* (ed. R. Milkman), pp. 60–82. Massachusetts: Sunderland.
—— (1983). Microevolution and macroevolution. In *Evolution from Molecules to Men* (ed. D. S. Bendall), pp. 387–402. University Press, Cambridge.
Beatty, J. (1982). Classes and cladists. *Syst. Zool.* **31**, 25–34.
Bishop, M. J. (1982). Criteria for the determination of the direction of character state changes. *Zool. J. Linn. Soc.* **74**, 197–206.
Bock, W. J. (1981). Functional-adaptive analysis in evolutionary classification. *Am. Zool.* **21**, 5–20.
Bookstein, F. L. (1983). Comments on a 'nonequilibrium' approach to evolution. *Syst. Zool.* **32**, 291–300.
Bremer, K. and Wanntorp, H–E. (1979). Hierarchy and reticulation in systematics. *Syst. Zool.* **28**, 624–7.
Brooks, D. R. and Wiley, E. O. (1984). Evolution as an entropic phenomenon. In *Evolutionary Theory: Paths into the Future* (ed. J. W. Pollard), pp. 141–72. John Wiley, Chichester.
Bush, G. L., Case, S. M.. Wilson, A. C., and Patton, J. L. (1977). Rapid speciation and chromosomal evolution in mammals. *Proc. Nat. Acad. Sci. USA,* **74**, 3942–6.
Cain, A. J. and Harrison, G. A. (1960). Phyletic weighting. *Proc. Zool. Soc. Lond.* **135**, 1–31.
Charig, A. J. (1982). Systematics in biology: a fundamental comparison of some major schools of thought. In *Problems of Phylogenetic Analysis* (eds K. A. Joysey and A. E. Friday), pp. 362–440. Academic Press, London and New York.
Charlesworth, B., Lande, R., and Slatkin, M. (1981). A neo-Darwinian commentary on macroevolution. *Evolution,* **36**, 474–98.
Cracraft, J. (1974). Phylogenetic models and classification. *Syst. Zool.* **23**, 71–90.
—— (1981). The use of functional and adaptive criteria in phylogenetic systematics. *Am. Zool.* **21**, 21–36.
Crowson, R. A. (1970). *Classification and Biology.* Heinemann, London.
D'Arcy Thompson, W. (1942). *On Growth and Form.* University Press, Cambridge.
Dover, G. A. and Flavell, R. B. (eds) (1982). *Genome Evolution.* Academic Press, London and New York.
Edwards, A. W. F. (1972). *Likelihood.* University Press, Cambridge.
Eldredge, N. and Cracraft, J. (1980). *Phylogenetic Patterns and the Evolutionary Process.* Columbia University Press, New York.
—— and Gould, S. J. (1972) Punctuated equilibria: an alternative to phyletic gradualism. In *Models in Paleobiology* (ed. T. J. M. Schopf), pp. 82–115. Freeman Cooper, San Francisco.

Farris, J. S. (1979). The information content of the phylogenetic system. *Syst. Zool.* **28,** 483–519.

—— (1983). The logical basis of phylogenetic analysis. In *Advances in Cladistics II* (eds V. Funk and N. Platnick), pp. 7–36. Columbia University Press, New York.

Felsenstein, J. (1978). Cases in which parsimony or compatibility methods will be positively misleading. *Syst. Zool.* **27,** 401–10.

—— (1983) Parsimony in systematics: biological and statistical issues. *Ann. Rev. Ecol. System.* **14,** 313–33.

Forey, P. J. (1982). Neontological analysis versus palaeontological stories. In *Problems of Phylogenetic Reconstruction* (eds K. A. Joysey and A. E. Friday), pp. 119–57. Academic Press, London and New York.

Fortey, R. A. and Jefferies, R. P. S. (1982). Fossils and phylogeny: a compromise approach. In *Problems of Phylogenetic Reconstruction* (eds K. A. Joysey and A. E. Friday), pp. 197–234. Academic Press, London and New York.

Friday, A. E. (1982). Parsimony, simplicity and what really happened. *Zool. J. Linn. Soc.* **74,** 329–35.

Ghiselin, M. T. (1974). A radical solution to the species problem. *Syst. Zool.* **23,** 536–44.

—— (1984). Narrow approaches to phylogeny: a review of nine books of cladism. *Oxford Surv. Evolut. Biol.* **1,** 209–22.

Gilmour, J. S. L. (1961). Taxonomy. In *Contemporary Botanical Thought* (eds A. M. MacLeod and I. S. Cobley), pp. 27–45. Oliver and Boyd, Edinburgh.

Gingerich, P. D. (1979). The stratophenetic approach to phylogeny reconstruction in vertebrate paleontology. In *Phylogenetic Analysis and Paleontology* (eds J. Cracraft and N. Eldredge), pp. 41–77. Columbia University Press, New York.

Goodman, M. (ed.) (1982). *Macromolecular Sequences in Systematic and Evolutionary Biology*. Plenum, New York and London.

Goodwin, B. C. (1984). Changing from an evolutionary to a generative paradigm in biology. In *Evolutionary Theory: Paths into the Future* (ed. J. W. Pollard), pp. 99–120. John Wiley, Chichester and New York.

——, Holder, N. J., and Wylie, C. C. (eds) (1983). *Development and Evolution*. University Press, Cambridge.

Gosliner, T. M. and Ghiselin, M. T. (1984). Parallel evolution in opisthobranch gastropods and its implications for phylogenetic methodology. *Syst. Zool.* **33,** 255–74.

Gould, S. J. (1980). Is a new and general theory of evolution emerging? *Paleobiol.* **6,** 119–30.

—— (1982). G. G. Simpson, paleontology and the modern synthesis. In *The Evolutionary Synthesis* (eds E. Mayr and W. B. Provine), pp. 153–72. Harvard University Press, Cambridge, Massachusetts.

—— and Eldredge, N. (1977). Punctuated equilibria: the tempo and mode of evolution reconsidered. *Paleobiol.* **3,** 115–51.

—— and Lewontin, R. C. (1979). The spandrels of San Marco and the Panglossian paradigm: a critique of the adaptationist programme. *Proc. Roy. Soc. Lond.* **B205,** 581–98.

Harper, C. W. (1976). Phylogenetic inference in paleontology. *J. Paleontol.* **50,** 180–93.

Harré, R. (1972). *Philosophies of Science*. University Press, Oxford.

Hecht, M. K. and Edwards, J. L. (1977). The methodology of phylogentic inference above the species level. In *Major Patterns in Vertebrate Evolution* (eds M. K. Hecht, P. C. Goody, and B. M. Hecht), pp. 3–51. Plenum, New York and London.

Hennig, W. (1966). *Phylogenetic Systematics*. University of Illonois Press, Chicago and London.

Hoffman, A. (1982). Punctuated versus gradual mode of evolution. *Evolut. Biol.* **15**, 411–36.

Hull, D. (1970). Contemporary systematic philosophies. *Ann. Rev. Ecol. System.* **1**, 19–53.

—— (1976). Are species really individuals? *Syst. Zool.* **25**, 174–91.

—— (1979). The limits of cladism. *Syst. Zool.* **28**, 416–40.

Kimura, M. (1983). *The Neutral Theory of Molecular Evolution*. University Press, Cambridge.

King, J. L. and Jukes, T. H. (1969). Non-Darwinian evolution. *Science,* **164**, 788–98.

Kitts, D. B. (1974). Paleontology and evolutionary theory. *Evolution,* **28**, 458–72.

Levinton, J. S. (1983). Stasis in progress: the empirical basis of macroevolution. *Ann. Rev. Ecol. System.* **14**, 103–37.

Løvtrup, S. (1975). A reexamination of the arachnid theory on the origin of the Vertebrata. *Zoologica Scripta,* **4**, 53–8.

—— (1977). *The Phylogeny of the Vertebrata*. John Wiley, London.

—— (1983). Victims of ambition: comments on the Wiley and Brooks approach to evolution. *Syst. Zool.* **32**, 90–6.

McDonald, J. F. (1983). The molecular basis of adaptation: a critical review of relevant ideas and observations. *Ann. Rev. Ecol. System.* **14**, 77–102.

McNeill, J. (1982). Phylogenetic reconstruction and phenetic taxonomy. *Zool. J. Linn. Soc.* **74**, 337–44.

Maynard Smith, J. (1983). Current controversies in evolutionary biology. In *Dimensions of Darwinism* (ed. M. Grene), pp. 273–86. University Press, Cambridge.

Mayr, E. (1981). Biological classification: towards a synthesis of opposing methodologies. *Science,* **214**, 510–16.

—— '(1982). *The Growth of Biological Thought: Diversity, Evolution, and Inheritance*. Harvard University Press, Cambridge, Massachusetts.

Nelson, G. (1984). Systematics and evolution. In *Beyond Neo-Darwinism* (eds M.-W. Ho and P. T. Saunders), pp. 143–58. Academic Press, London and New York.

—— and Platnick, N. (1981). *Systematics and Biogeography*. Columbia University Press, New York.

O'Grady, R. T. (1984). Evolutionary theory and teleology. *J. theoret. Biol.* **107**, 563–78.

Panchen, A. L. (1982). The use of parsimony in testing phylogenetic hypotheses. *Zool. J. Linn. Soc.* **74**, 305–28.

Patterson, C. (1977). The contribution of paleontology to teleostean phylogeny. In *Major Patterns in Vertebrate Evolution* (eds M. K. Hecht, P. C. Goody, and B. M. Hecht), pp. 579–643. Plenum, New York.

—— (1981). Significance of fossils in determining evolutionary relationships. *Ann. Rev. Ecol. System.* **12,** 195–223.

—— (1982). Classes and cladists or individuals and evolution. *Syst. Zool.* **31,** 284–6.

Platnick, N. I. (1979). Philosophy and the transformation of cladism. *Syst. Zool.* **28,** 537–46.

Rachootin, S. P. and Thomson, K. S. (1981). Epigenetics, paleontology, and evolution. In *Evolution Today* (eds G. C. E. Scudder and J. L. Reveal). *Proceedings of the Second Internationl Congress of Systematics and Evolutionary Biology*, pp. 181–93.

Raff, R. A. and Kaufman, T. C. (1983). *Embryos, Genes, and Evolution.* Macmillan, New York and London.

Reif, W-E. (1983). The German paleontological and morphological tradition. *In Dimensions of Darwinism* (ed. M. Grene), pp. 173–203. Cambridge University Press, Cambridge.

Rensch, B. (1959). *Evolution above the Species Level.* Methuen, London.

Riedl, R. (1978). *Order in Living Organisms.* John Wiley, Chichester and New York.

Rudwick, M. J. S. (1964). The inference of function from structure in fossils. *Br. J. Philos. Sci.* **1,** 27–40.

Schaeffer, B., Hecht, M. K., and Eldredge, N. (1972). Phylogeny and palaeontology. *Evolution. Biol.* **6,** 31–46.

Simpson, G. G. (1944). *Tempo and Mode in Evolution.* Columbia University Press, New York.

—— (1953). *The Major Features of Evolution.* Columbia University Press, New York.

Sneath, P. H. A. and Sokal, R. R. (1973). *Numerical Taxonomy.* Freeman, San Francisco.

Sober, E. (1983). Parsimony in systematics: philosophical issues. *Ann. Rev. Ecol. System.* **14,** 335–57.

Stanley, S. M. (1975). A theory of evolution above the species level. *Proc. Nat. Acad. Sci. USA,* **72,** 646–50.

—— (1979). *Macroevolution, Pattern and Process.* Freeman, San Francisco.

—— (1981). *The New Evolutionary Timetable.* Basic Books, New York.

Vbra, E. S. (1984). Patterns in the fossil record and evolutionary processes. In *Beyond Neo-Darwinism* (eds M-W. Ho and P. T. Saunders), pp. 115–42. Academic Press, London and New York.

—— and Eldredge, N. (1984). Individuals, hierarchies and processes: towards a more complete evolutionary theory. *Paleobiol.* **10,** 146–71.

Webster, G. (1984). The relations of natural forms. In *Beyond Neo-Darwinism* (eds M-W. Ho. and P. T. Saunders), pp. 193–217. Academic Press, London and New York.

Wicken, J. S. (1981). Evolutionary self-organisation and the entropy principle. *Nature and System,* **3,** 129–42.

—— (1983). Entropy, information, and nonequilibrium evolution. *Syst. Zool.* **32,** 438–43.

—— (1984). On the increase in complexity in evolution. In *Beyond Neo-Darwinism* (eds M-W. Ho and P. T. Saunders), pp. 89–112. Academic Press, London and New York.

Wiley, E. O. and Brooks, D. (1982). Victims of history-a non-equilibrium approach to evolution. *Syst. Zool.* **31,** 1–24.

Fisher's evolutionary faith and the challenge of mimicry

JOHN R. G. TURNER

Dedicated to the memory of Bernard Norton (d. 1984) – a loss to his friends and to the historical study of modern biology.

'. . . not the wing of a butterfly can change in form, or vary in colour, except in harmony with, and as a part of, the grand march of nature'.

Wallace (1865)

'That nothing walks with aimless feet'

A random, purposeless world is not a comfortable world to live in. A world of order and purpose, in which we can trust that '*somehow*, good shall be the final goal of ill', can give us a sense of our own purpose. Tennyson tottered on the brink of the alternative world when he wrote that pregnant 'somehow' in *In Memoriam*, 20 years before the *Origin of Species*. After the Darwinian event, with its revelation of suffering as the driving force of evolution, despair about the universe became a real option.

> We are no other than a moving row
> Of visionary Shapes that come and go
>   Round with this Sun-illumin'd Lantern held
> In Midnight by the Master of the Show;
>
> Impotent Pieces of the Game he plays
> Upon his Checker-board of Nights and Days;
>   Hither and thither moves, and checks, and slays;
> And one by one back in the Closet lays.
>
> When You and I behind the Veil are past,
> Oh but the long long while the World shall last,
>   Which of our Coming and Departure heeds
> As much as Ocean of a pebble-cast.
>
> And that inverted Bowl we call The Sky,
> Whereunder crawling coop'd we live and die,
>   Lift not your hands to *It* for help – for It
> As impotently rolls as you or I.

The further suggestion, from the Second Law of Thermodynamics, formulated toward the end of the century (Brush 1976), that however long a while the world might last, it would eventually run itself down to just-above-absolute-zero entropy, could cause even the least Tennysonian temperament a moment or two of disquiet.

Many solutions have been tried (Bowler 1983, 1984): Lamarckism, Vitalism, even outright Faith; some in America have believed that breaking the bloody glass would hold up the weather and (while not actually breaking it) have poured the contents over E. O. Wilson's head. Among these solutions, all of them ambivalent (just why Lamarckian inheritance takes the serpent's tooth from the cruelty of the world is by no means immediately obvious), none is more ambivalent than natural selection. Creationists and even physicists insist that we could not have evolved from amoebas by 'blind chance' (DuNouy 1947; Hoyle & Wickramasinghe 1981). In vain do biologists point out that natural selection is a system for generating very low states of probability (Fisher 1954; they are being purely pedantic if they quibble about the amoeba): for the believing creationist, natural selection is as blind and as chancy as mutation. It lacks purpose. It is, what is more, wasteful and cruel.

> Pain, grief, disease, and death, are these the inventions of a loving God? That no animal shall rise to excellence except by being fatal to the life of others, is this the law of a kind Creator? It is useless to say that pain has its benevolence, that massacre has its mercy. Why is it so ordained that bad should be the raw material of good? Pain is not the less pain because it is useful; murder is not the less murder because it is conducive to development. Here is blood upon the hand still, and all the perfumes of Arabia will not sweeten it (Winwood Reade's *Martyrdom of Man*, quoted by Wallace, 1889, pp. 36–7).

(The serpent's tooth extracted by Lamarckism is the *necessity* of cruelty; with Lamarckian evolution, cruelty and waste are only contingent excrescences on the world – God is merely incompetent; with Darwinism, they are the prior necessities for our own existence – God is Manichean.) .

Yet, natural selection, on top of being the only adequate materialist explanation so far devised for the fact that organisms are adapted to the world in ways that rocks are not (Dawkins 1983) can be seen also as the direct working of a beneficial creative force. There is something very beautiful about adaptive devices, and maybe a crumb of subconscious comfort in the thought that the world could have endowed its creatures with such armoury against the slings and arrows of outrageous fortune. The principle of natural selection comes trailing clouds of glory from a former and more comforting philosophy: adaptation could once be seen as proof of the wisdom and beneficence of a designing creator; it can now be seen as evidence of a process that, even if not merely 'careless of the single life', but rather sacrificing it in the name of progress, is in its own twisted way, beneficent. The structure of European languages (perhaps of all human languages) is teleological. Without the most tortuous circumlocutions, we cannot describe adaptation and selection without implying purpose. Although most of us maintain that this is metaphor, which in strict logic it is, at the emotional level the liturgy about beneficent and goal-directed acts must be comforting.

The seeking of optimistic alternatives to Darwinism (as with Butler and Shaw), or a coupling of Darwinism, agnosticism and, if one's temperament ran to it, gentle Omarian despair, seem to have been the two ways chosen by many intellectual late-Victorians. However, if one is a rigorous biologist, neo-Lamarckism will not do; neither, if one is a Christian, will

Omarism. What follows is the saga of a man who sought resolution in the optimistic view of natural selection; who, against the contemporary odds, saw Darwin's theory as implying the order and purpose attributable to a Christian Creator. For among those biologists who have wanted to see natural selection in this beneficent way was R. A. Fisher. His daughter tells us that he was born, lived, and died in the Anglican faith (Fisher Box 1978). He ardently hoped for human progress through the practice of eugenics. Through his filial relationship with the childless Leonard Darwin, who gave him encouragement, help, even money (Bennett 1983), he became in all but the legal sense, the adopted grandson of Charles Darwin himself. This triad – his Anglican faith, his hopes for the eugenic future of mankind, and his role as emotional and intellectual heir of Charles Darwin – played intimately within itself in the development of Fisher's thought. It is beyond my scope to speculate which event was at which point of his life, the leader. But the programme he eventually developed shows just how tough-minded he was: the reconciliation of any Christian faith, other than the dourest Calvinism – in which God the father truly is the Godfather – with the world of strict Darwinism was one of the most unpromising compromises one could attempt (Bowler 1983, 1984). No doubt the optimism inherent in eugenics was one of the compensations.

> To the traditionally religious man, the essential novelty introduced by the theory of . . . evolution . . . , is that creation was not all finished a long while ago, but is still in progress . . . In the language of Genesis we are living in the sixth day, probably rather early in the morning . . .
> . . . the effective causes of evolutionary progress lie in the day-to-day incidents in the innumerable lives of innumerable plants, animals and men. We need not think of the effective causes as . . . imposed upon the organic world by the external environment; but where the organism meets its environment, where it succeeds or fails in its endeavours, where its potentialities are tried out in practice and incarnated in real happenings, there it is that the doctrine of natural selection locates the creative process (Fisher 1947).

This picture of natural selection was a favourite theme of Fisher's; it appeared, with very similar wording, in his first paper on the subject (1934), and again in his Eddington Lecture (1950):

> Just where does the theory of natural selection place the creative causes which shape evolutionary change? In the actual life of living things; in their contacts and conflicts with the environment . . .; in their unconscious efforts to grow, or their more conscious efforts to move. Especially, in the vital drama of the success or failure of each of their enterprises (p. 17).

The future was, Fisher believed, indeterminate, or rather determined only in the statistical sense – for otherwise we would know the future equally as clearly as the past. The future was truly being created, in that only one of a range of possible futures was actually realized. It was there that Fisher located our Free Will – which he had accepted as an article of belief since his undergraduate days at least (Fisher 1912) – and its interaction with the Will of the Creator. In this interaction 'living things themselves are the chief instruments of the Creative activity' (1950), for their decisions, freely made, affect their probabilities of survival, death, and reproduction, which are themselves the agencies of natural selection. It is thus through natural selection, an interaction between willing Creator and freely choosing Creature, that God, in Charles Kingsley's words, 'makes things make themselves'. For Fisher, the nexus could encompass the problem of evil – which had something to do with wrong choices; where neo-Lamarckism supposed that mere willing was sufficient, Darwinism entailed the Christian values not only of Faith but, in 'doing or dying', of Works.

Fisher, the premier statistician of his generation, committed, so it seems, to this statistical view of causality that encompassed true freedom of choice, was yet firmly of the opinion that what the rest of the world saw as the two stochastic processes of genetics – mutation and random genetic drift – had no significant influence on the direction of evolution. It looks like a hiatus in Fisher's thought – Dr Jekyll, the stochastic theorist, becomes Mr Hyde, the old fashioned, unenlightened determinist (Allen 1983). In modern demonology, to be a 'determinist' fits very well with Fisher's being something equally evil – a eugenist!

History and Fisher is more subtle than that. I shall argue that the purely stochastic processes of thermodynamic chaos, to which the molecular disordering of the genetic material – mutation – and the disordering of the genetic structure of populations – random drift – both belong, can be seen not merely as non-determinate. They are arbitrary and degenerative. For this reason, Fisher saw them as having no large or important part in evolution, a process that was constructive and progressive.

Hence, Fisher's first task had been to defeat the ascendant Mutationist school, who saw a significant creative role for the mutational process itself. No sooner had he won that battle, than he was confronted by the allegation that random genetic drift could also be creative. His great book *The Genetical Theory of Natural Selection* (1930, hereafter referred to as the *Genetical Theory*) stands at the end of the first battle, and at beginning of the second. The concepts which Fisher developed in the two conflicts have informed much of the modern theory of evolution, as it is understood by the more technically-minded biologist, particularly in England. What may be fairly called the Oxford School of Ecological Genetics, represented at the international level by E. B. Ford, H. B. D. Kettlewell, A. J. Cain, P. M. Sheppard, and B. C. Clarke, developed their studies on natural selection, random drift, and mimicry within a framework that was deeply influenced by the scientific views of Fisher, who was at that time resident in England's other ancient and dominant university at Cambridge.

Fisher then fought two battles against arbitrariness and chaos, from which grew what was almost a distinctive national style in population genetics. This paper recounts the history and consequences of Fisher's first battle, against Mutationism, between the 1920s and 1960s. Its partner (Turner 1985) deals with the second battle, the response of Fisher and the Oxford School to the question of random drift. With the wisdom of hindsight, we can see that in both the problem was one of striking a balance between the adaptive and non-adaptive forces of evolution: a singularly tricky piece of mental acrobatics. In the first battle, the problem became focussed, a little improbably, on the exotic question of mimicry in butterflies. (See also, for further aspects of this history, Kimler 1983a,b; Turner 1983, 1986.)

'The dangerous fallacy'

Mimicry at first sight is indeed an odd and off-beat subject for Fisher to choose in developing a theory of evolution. He was primarily a geneticist, statistician, mathematical theorist, and eugenist, not a naturalist or bug-hunter, and we now have a tradition that regards any work on evolutionary genetics conducted with anything larger than *Drosophila* as impractical, if not a bit of a joke (Lewontin 1972). This attitude did not exist anything like so strongly in the 1920s. *Drosophila* was being developed as a genetic tool, but in the first two decades of the century much less compact organisms had been the normal tools, and various butterflies and moths had figured prominently, a tradition which was to be continued, against the trend, by both the Oxford School and by Goldschmidt (1945). Even within this tradition, Goldschmidt and Fisher were exceptional in treating the question of mimicry as central, rather than in its usual role as a pretty, but peripheral example of adaptation.

To see why Fisher needed to devote a whole chapter of the *Genetical Theory* to mimicry and to emphasize the point with two colour plates, we need to see how mimicry had come to pose the major challenge to Fisher in his attempts to cut mutationism down to size.

The conflict between mutationists (or saltationists) and Darwinists during the first 20 years of this century (Provine 1971) can be seen as a dispute over creation. The Darwinists of the late nineteenth century tended to see natural selection as creating, in a true sense of the word, the adaptations of organisms from the almost plastic, continuous variation which by then was known to be a universal feature of natural populations (Wallace 1889). As Punnett (1915) wryly remarked, there was no large change in scientific viewpoint from the pre-Darwinian one: Natural Selection had simply replaced God in the litanies of adaptationist praise. The mutationists had taken a more modern view of creation, in seeing it as the selection of adaptive forms from a whole array of forms, generated by the process of genetic mutation. 'The creature' said Bateson, 'is beheld to be very good after, not before its creation'.

This view gives a truly random, non-adaptive process a part to play in creation. The mutationists assumed that the elements which would be fed to the selective machinery were quite large, right up to the whole of the difference between two species. In a sense, the creating had been done by the mutational, rather than by the selective process. The modern view is that this is a false antithesis and that creation, perhaps all creation, even in the human mind consists of both processes: the selection of the most suitable few from a range of alternatives generated more or less at random.

> . . . we cannot regard mutation as a cause likely by itself to cause large changes in a species. But I am not suggesting for a moment that selection alone can have

any effect at all. The material on which selection acts must be supplied by mutation. Neither of these processes alone can furnish a basis for prolonged evolution (Haldane 1932, p. 110).

Punnett had gone a long way toward this view as early as 1915, by proposing that what natural selection did was to propel gene mutations into the population.

> . . . a rare sport is not swamped by intercrossing with the normal form, but . . . on the contrary if it possess even a slight advantage, it must rapidly displace the form from which it sprang (Punnett 1915, p. 143).

In trying to redress the balance against the late-Darwinian view of an all-powerful natural selection, despite his clear belief that evolution consisted of the selection of mutations, Punnett consistently emphasized that it was mutation which did the creating:

> . . . the function of natural selection is selection and not creation (Punnett 1911, p. 132).

'All is flux'

It was this position, what his colleague Ford was later to call 'the dangerous fallacy of mutation', that Fisher sought to reverse. He had available for his intellectual mill the findings of 'the second mutation theory' (Olby 1981). The increasing sophistication of genetic analysis had confirmed, during the later teens of the century, the Darwinists' suspicions that not all mutations were monstrosities, in which considerable changes could be wrought on the phenotype: many were small enough to give rise to the continuous variation which the Darwinists had supposed to be the material moulded by natural selection.

Fisher, having put the final touches to showing that the finer points of the inheritance of continuously varying characters could be explained and predicted by numerous Mendelian genes each of individually very small effect (Fisher 1918), and hence having finally proved what had long been suspected, that the biometrical and Mendelian views on heredity were not antithetical, set out to show that, likewise, the Mendelian and Darwinian views of evolution, as he saw them, could be reconciled. He seems to have set himself this task early in his career, perhaps even as an undergraduate:

> I first came to Cambridge in 1909, the year in which the centenary of Darwin's birth and the jubilee of . . . *The Origin of Species* were being celebrated. The new school of geneticists using Mendel's laws . . . was full of activity and confidence, and the shops were full of books good and bad from which one could see how completely many writers of this movement believed that Darwin's position had been discredited (Fisher 1947).

The work of synthesis had commenced, with a clear eye on the human implications, with the two papers which he read to undergraduate meetings of the Cambridge Eugenics Society in 1911 and 1912 [reprinted in Bennett

(1983); this compilation, with its valuable intellectual biography, is hereafter called *Letters*]. He was to look back on the final achievement with justified satisfaction:

> As a tradition . . . genetics is exposed more indefensibly than you seem to admit to the criticism of being anti-Darwinian . . . in . . . factiously attacking and trying to discredit the far-reaching and penetrating ideas on the *means* of organic evolution which Darwin had originated. It was not only Bateson and de Vries, but almost the whole set of geneticists in the first quarter of this century, who discredited themselves in this way . . . Writer after writer asserted . . . that species arose by single mutations, and that selection of small continuous variations within the species was known to be inoperative pending the arrival of an appropriate mutation . . . The idea of polygenic Mendelism was frowned upon . . . [The] assertion . . . that particulate inheritance, so far from being antagonistic to Darwin's main theory, actually removed the principal difficulty with which it was encumbered . . . was entirely new when I put it forward in 1930 (Fisher to K. Mather, 1943, *Letters*, p. 236).

What Fisher is maintaining is not, we trust, that nobody understood the way in which natural selection would cause evolutionary changes by altering the frequency of mutations in populations. That understanding had commenced in Punnett's *Mimicry in Butterflies* of 1915, particularly in H. T. J. Norton's mathematical appendix, and by 1930 Haldane had developed it into a sophisticated theory, in his series of papers for the Cambridge Philosophical Society. It is now the mainstay of the elementary theory of population genetics. Fisher maintained that, before him, nobody had understood how continuous, gradual evolution, using not newly-arising mutations, but the already existing continuous variation exhibited by the population – the kind of evolution postulated by the late-Victorian Darwinians (I leave for other times and other people the question of Darwin's own views) – could be explained as a result of Mendelian inheritance.

It is a claim that will stand up at least to superficial scrutiny. Punnett had succumbed on the question of continuous variation being explicable by Mendelian genes as early as the 1919 edition of the text-book *Mendelism*, but even the last edition of 1927 continued to assert the view outlined by Fisher in the above letter, that continuous variation was non-heritable 'fluctuation' which could play no part in evolution.

Having devoted the first chapter of the *Genetical Theory* to showing how the standing variation of a population, if it was Mendelian, would not be drained away by blending inheritance – the 'principal difficulty' of Darwin's theory according to the letter and to much modern historical opinion – Fisher set out to produce a dynamic theory of evolution resulting from the action of natural selection on such variation. His training in physics and statistics led him to seek general principles and a global picture, rather than the intimate details of the changes in means and variances of characters, that have been worked out since. The genes in the population, he saw, could be treated in much the same way as the molecules in a gas.

New mutations of small effect would be continuously occurring. Many were lost stochastically not long after they appeared; a minority of those which were advantageous or close to neutrality in their effects on fitness would increase stochastically in frequency to form a dynamic pool of variation in the population, and from this pool those which were, or became, advantageous would start a steady increase in their frequency, eventually, if their advantage was not reversed, reaching close to 100 per cent. The advantageous mutants that were passing through the population in this way constituted a perpetual flux of evolutionary change, perpetual because the environment constantly changes and the organism can never be perfectly adapted. Fisher was not an adaptationist in the naive sense of believing that all parts of all organisms are, like the House of Peers 'not susceptible of any improvement at all'. It is their very imperfection that makes some mutations beneficial.

> . . . we can conceive, though *we need not expect to find*, biological populations in which the genetic variance is absolutely zero, and in which fitness does not increase (*Genetical Theory*, p. 36, italics added). [P. 39 in the 1958 reprint – page numbers from this edition will be given in square brackets throughout.]

Fisher developed two crucial arguments, as we shall see – the continual flux of evolutionary change and its consisting only of mutations of small effect – from this picture of the population perpetually failing in its attempts to catch the point of maximum adaptedness, like Browning's lover pursuing his unattainable mistress:

> No sooner the old hope goes to ground
> Than a new one, straight to the self-same mark
> I shape me –
> Ever
> Removed!

If a gene is increasing in frequency as it passes through the flux, then *ex hypothesi* it is, on average, fitter than the average for the whole population. Similarly, genes which are declining in the number of copies by which they are represented are less fit than average. If we take the extent by which all the individuals who possess a particular gene are raised above, or depressed below, the average fitness of the whole population, then we have a measure which Fisher called the average excess (a) by which the gene exceeds (or falls short of) the mean for the whole population. It is now possible to calculate a similar deviation for the average offspring of each individual who carries the gene in question; Fisher called this the average effect (α) of the gene, and showed that the variance in fitness of the whole population could be calculated as

$$V = 2\Sigma_i q_i a_i \alpha_i$$

where q_i is the frequency at the present moment of the ith gene, and a_i and α_i are its average effect and average excess (*Genetical Theory*, p. 33 in both editions, but incorrect in the 1930 edition). This is a mathematical distillation, in the precisely definable concept of variance, of the fact that in an evolving population there must be genetic variation in the fitness of individuals.

However, if the fitter individuals are having more offspring than average, and hence the fitter genes are becoming commoner in the population, it follows that the average fitness of the whole population must be increasing. The moment to moment increase in the average fitness was, Fisher showed, equal to V. This was the Fundamental Theorem of Natural Selection:

> *The rate of increase in fitness of any organism at any time is equal to its genetic variance in fitness at that time'* (*Genetical Theory*, p. 35 [p. 37]).

As the variance must always be positive (this is a necessary property of variances), it followed that fitness was always increasing and would do so at a rate which was governed by the amount of beneficial variation passing through the population. Fisher had therefore found a parameter which related the speed of evolution directly to the available variation. That was an important achievement, but Fisher was seeking even bigger fish.

> It will be noticed that the fundamental theorem . . . bears some remarkable resemblances to the second law of thermodynamics. Both are properties of populations, or aggregates, true irrespective of the nature of the units which compose them; both are statistical laws; each requires the constant increase of a measurable quantity, in the one case the entropy of a physical system and in the other the fitness . . . of a biological population . . . Professor Eddington has recently remarked that 'The . . . second law of thermodynamics . . . holds, I think, the supreme position among the laws of nature'. It is not a little instructive that so similar a law should hold the supreme position among the biological sciences (*Genetical Theory*, pp. 36–7 [p. 39]).

Both might, Fisher suggested, 'ultimately be absorbed by some more general principle', although there were some profound differences: entropy increase occurred in a universe which was permanent, fitness increase in populations liable to extinction; 'fitness, although measured by a uniform method, is qualitatively different for every different organism' – entropy was the same for all physical systems; fitness was subject to change by the external action of the environment; and entropy changes were exceptional among physical processes in being irreversible, while the irreversible increase of fitness was only one of the many irreversible processes in biology.

> Finally, . . . entropy changes lead to a progressive disorganisation of the physical world, at least from the human standpoint of the utilization of energy, while evolutionary changes are generally recognized as producing progressively higher organization in the organic world (*Genetical Theory*, p. 37 [p. 40]).

Here, perhaps, we have the central emotional appeal of the Fundamental Theorem (and I am entirely indebted to my friend Jonathan Hodge for this insight). The Second Law and the Fundamental Theorem are Yin and Yang – possibly both capable of being combined into one Tao – 'some more general principle'. While conforming mathematically, they lead to opposite ends: the one to the heat death of the universe (or perhaps, Fisher hopes, only to an inconvenient loss of usable energy), the other to the improved adaptation of organisms. Fisher liked to see himself as tough-minded and disliked the more morbid styles of romantic poetry typified by the Nightmare Life-in-Death of the *Ancient Mariner*:

> Horror, disgust, superstitious terror are emotions familiar enough to the human race. Are they *worth* all this screaming emphasis? (Fisher to Nora Barlow, 1958, *Letters*, p. 182).

He is not to be found bewailing the individual sacrifices required by the doing and dying of natural selection: where the individual met its environment, and succeeded or failed in its endeavours, there it was that the theory of natural selection located the creative process. Not a worm was cloven in vain. Death was swallowed up in adaptation.

What was this 'fitness' that was in a state of beneficial, but in a sense perpetually frustrated, increase? A strict mathematical interpretation is that it is the rate of exponential population growth or, in a population subject to density regulation, the size of the standing population. But other commentators (Grene 1974; Olby 1981) have perceived that something more was intended:

> 'Fitness, although measured by a uniform method, is qualitatively different for every different organism . . .' (*Genetical Theory*, p. 37 [p. 39]).

Here different 'organism' appears to mean 'species', as in the Fundamental Theorem, and the 'fitness' to be some property, roughly what we would now call competitive ability, of the whole population.

Fisher noted that, although the quality of the organism was perpetually increasing, the increase was just as perpetually offset by 'deterioration' of the environment – deterioration that is from the point of view of the adapted organism. This included both secular changes in the physical environment and adaptive evolutionary changes in other 'organisms' which were perpetually increasing their fitness. We would now say that the single species is, like the Red Queen, doing all the running it can to stay in the same place, and that different species which prey on each other or compete with each other, are running an evolutionary arms race.

Those 'organisms' which could not keep up in the race would, according to Fisher, become extinct; the chief risk, he thought, was to have a small population size, and therefore to lack a sufficient supply of mutations to feed into the flux, upon which the increasing fitness depended. From there, Fisher went on to explain a major feature of human history: the decay of civilizations. The most valued members of civilized societies were socially promoted into a class where, by the restraints of wealth and inheritance, they were induced to have smaller families. In conflict with barbarian societies, who rewarded their most valued members with more children, the civilized societies therefore failed to increase their competitive ability fast enough, and succumbed (*Genetical Theory*, Chapter XI). Civilized societies, in short, had perverted the natural process described by the Fundamental Theorem.

Fisher's programme in eugenics, therefore, was not to build the once and for all perfect human being – the point of maximum adaptation could never be reached – but merely to prevent this decline of civilizations by restoring the natural process within their societies or, if you prefer, putting them on an equal footing with barbarians, by ensuring that the most valued members of society – that is those who had been socially promoted – did again have the larger families. Family allowances scaled in proportion to taxable income would, he thought, do the trick. (Note how closely, and beautifully Fisher's science and eugenics are joined together and inform each other. It would be hard to assert that the eugenic proposals were merely a lineal deduction from the scientific theories; see also Norton 1983.)

Evolution was therefore reaching the point where the percipient will of the perpetually perfectable creature could take over part of the creative role.

> I think . . . that we must regard the human race as now becoming responsible for the guidance of the evolutionary process acting upon itself (Fisher to P. de Hevesy, 1945, *Letters*, p. 192).

> Man is in process of creation, and the process involves something we can call improvement, in which Man's own co-operation is necessary (Fisher to Bishop E. W. Barnes, 1952, *Letters*, p. 182).

> . . . the Divine Artist has not yet stood back from his work, and declared it to be 'very good.' Perhaps that can be only when God's very imperfect image has become more competent to manage the affairs of the planet of which he is in control (Fisher 1947).

The inverted bowl we call the sky was impotent to deliver help into our uplifted hands, but, Fisher might have retorted impatiently, the maudlin romantic poet had failed to notice that our hands already held the means of help within their grasp.

'Small is beautiful'

Within this system of ultimately beneficial natural selection, Fisher saw no room for the forces of chaos and old night. A randomizing process may, as I have said, feed in the variation on which creative selection must work. Indeed, it is hard to see how creation is possible without it. The process which randomizes the biological material at the population level, random genetic drift, Fisher was later to dismiss completely (see, for example, Provine 1985; Turner 1985) and he paid no serious attention to it in the *Genetical Theory*. To mutation, randomization as we now know at the molecular level, he assigned the minimum possible creative role – and here 'minimum' is not inflated rhetoric for 'least', but signifies 'tending to zero', in the way that concept is used in calculus. Thus while it was clear that mutations come in all sizes – and we must remember that at that time 'size' was chiefly seen in terms of the effect on the phenotype, not in the number of base-pairs of DNA altered – it was only those which tended to zero which had a significant part to play in evolution. This left natural selection as the 'onlie begetter' of adaptation, and denied any creative role to stochastic processes.

With his commitment to freedom of will, indeterminacy, and statistical causation, why was Fisher so opposed to giving a large role to the stochastic forces of mutation and random drift? I hazard that it was because these introduce a large element not just of indeterminacy, but of arbitrary and meaningless chance. The indeterminacy of natural selection was different; not only was its global outcome, the increase of adaptation, predictable, beneficial, and in some way purposive, but its operation depended upon the action of the individual 'where it succeeded or failed in its endeavours'; the whole process was subject to change by the individual will and, with eugenic policy, by the general will also. It was in that interaction of the willing creature and the (in the long term and statistically and if we co-operate) beneficient Creator, that the 'doctrine of natural selection locate[d] the creative process'. No amount of will could do anything to affect mutation and drift, and as chaotic forces they represent only the Mephistophelean Second Law of Thermodynamics and 'der Geist, der stets verneint', 'the spirit that says *No*'.

I do not want to make too much of the religious influence on Fisher's thought. He was, his daughter tells us, 'always a scientist' (Fisher Box 1978) and no doubt would, and did, adjust his philosophy when the scientific facts demanded it. Such an adjustment must have taken place when he opted for Darwinism at a time when most optimists preferred neo-Lamarckism or other non-Darwinian theories (Bowler 1983). Fisher did have sound scientific reasons for rejecting mutationism and random drift. Mutationism, in the full-blown original version of Bateson or of de Vries provided no adequate explanation of adaptation, and therefore as science fared no better than creationism, or the divinely directed evolution of Asa Gray and the Duke of Argyll (Bowler 1983). But Fisher's rejection of the weaker form of mutationism, in which some evolutionary changes were large, was – as we shall see – not even internally consistent with his own evolutionary model. Random drift he likewise rejected on what was no more than a hunch about the size of natural populations and the rate of migration between them (Turner 1985). Evidence on either point was clearly not the crucial factor in persuading Fisher to take his personal viewpoint. This surely resulted from his adoption of the form of Darwinism – often appropriately called Wallacism – which saw selection as the *allmächtig* force of evolution, and his ability to see in its workings the hand of the Almighty. As he humorously observed (1950), while our loyalty to the facts must be absolute and the universe may turn out to be repugnant to our philosophical prejudices, to assume that the facts will necessarily be adverse to our hopes and aspirations, is downright perverse!

There is a world of difference between that which is free, in the sense of not being precisely determined, and that which is arbitrary, chaotic, and accidental; evolution by selection was in the first category, evolution by mutation and random drift in the second. The Fundamental Theorem, he said, allowed one to put in perspective 'the objection . . . that the principle of Natural Selection depends on a succession of favourable chances' (such as Shaw's characterization of it in *Back to Methuselah* as 'a chapter of accidents').

> The objection . . . depends for its force upon the ambiguity of the word chance, in its popular uses. The income derived from a Casino by its proprietor may, in one sense, be said to depend upon a succession of favourable chances, although the phrase contains a suggestion of improbability more appropriate to the hopes of the patrons of his establishment [a rare touch of humour this, in English academic dead-pan]. It is easy without any very profound logical analysis to perceive the difference between a succession of favourable deviations from the laws of chance, and on the other hand, the continuous and cumulative action of these laws. It is on the latter that the principle of Natural Selection relies (*Genetical Theory*, p. 37 [p. 40].

Evolution by mutation or by random drift would depend on a succession of favourable deviations from the laws of chance and would be so improbable as to be impossible. Natural selection, on the other hand, while 'requiring no rigid determinism whatsoever' (Fisher to C. S. Stock, 1936, *Letters*, p. 264) is anything but chaotic. The summation of its random, stochastic elements is a predictable, systematic trend, which will steadily generate very low states of probability (Fisher 1954). Selection was, *contra* Punnett, a creative force; mutation was not.

Fisher was not, obviously, in a position to deny that mutation was the raw material of evolutionary change (Olby 1981). How then was its 'creative' role to be denied? The mutationists had seen it as supplying the variation of organisms in large hunks to the selective machinery, even in the early extreme version in hunks large enough to constitute a new species. Fisher's solution was to assert that on the contrary the only important mutants in evolution were those whose effect on the phenotype was minimal – those which, in short, produced the standing continuous variation of populations. These could not be said to create adaptations, or organisms, or even parts of adaptations or organisms, any more than the particles in a lump of clay can be said to contribute to the form which is moulded by the hands of the sculptor. Such a thesis also met an outstanding criticism of the theory of evolution as it was then developing – and one that is still heard frequently from creationists – that the observed large mutations were hopelessly maladaptive monstrosities. Indeed, most of them were and that was why only small mutations, minimally small according to Fisher, played a part in evolution.

The beginning of Fisher's thoughts on this subject can be seen in a letter from Leonard Darwin of 1925 which states the problem and points the solution clearly enough.

> As to big mutations, I have no doubt they are generally harmful. But are not they rare and soon stamped out? If so, they are of no great importance in evolution. As to small mutations, these are what I believe evolution mainly relies on, and it seems to me difficult to prove that they are more often harmful than not . . . Perhaps there may be such a thing as an organism which is as perfectly adapted to its environment as selection can make it. In that case, *ex hypothesi*, every mutation must be harmful . . . (Darwin to Fisher, 1925, *Letters*, pp. 77–8).

But if an organism is not perfectly adapted, a very few mutations at any one time will be beneficial. Hence, Fisher's picture of the evolving organism in pursuit of the will o'the wisp, perfect adaptation. Now consider adaptation as a measurable function and imagine it as a mountain which the organism is trying to climb. It is easier if one thinks of it as a perfectly conical peak, like one of the Andean volcanoes. When the organism reaches the top it will be perfectly adapted. The problem for the organism is that it cannot 'know', in its genes, which way is 'up'. Mutation occurs in all directions, with no particular regard for improvement or otherwise. Moves from the current position therefore are made randomly in all directions. Which moves will produce improvements? If you are the organism, blindfolded and having no notion of the direction of the peak, should you try jumping or shuffling your feet? Jumping might seem to offer you the prospect of a more rapid improvement, but the bigger the jump, the less likely it is to take you actually up the hill, because the slope of the mountain curves away round the peak; more of the surrounding landscape is 'down' than 'up'. Your movement in traverse along the side of the mountain is rather likely to take you to a point which is actually lower than the one you started at. Only small shuffles of the feet give you a sporting 50:50 chance of going up or down. All more dramatic moves are weighted against you, and are more likely to move you down than up. To use a completely different analogy, if you have a tolerably well adjusted microscope or other piece of machinery, you may be able to improve its alignment by gently tapping various parts of it: kicking it seldom produces an improvement.

Therefore, Fisher argued – and the argument becomes much more impressive if the mountain is imagined in an infinite number of dimensions, which is more appropriate considering the large number of ways in which an organism must be adapted to its environment – almost all large mutations are disruptive. With, perhaps, some statistical exceptions, it will be only the mutations of very small effect that take part in adaptive evolution. The large mutations which Punnett and other mutationists had supposed to be the only stuff of evolution, had in fact no part to play whatever.

The outcomes of this theorem are three. First, there is the mathematical convenience that evolution can be described adequately by infinitesimal calculus, and the flux of mutations as a statistical fluid. Second, and concomitantly, evolution is, in a very real sense, gradual. Third, no significant creation is done by mutation: it feeds to the selective machine variation so finely divided as to be almost fluid, the particles of which it consists being not so much the minimum quanta of genetic variation (what we would now see as a single base pair change), but actually tending, in their phenotypic effects, statistically toward zero.

It is difficult to fault this argument as it stands, and it has greatly influenced most English population geneticists. But it does have the crucial premise that the 'organism' is within striking distance of only one point of maximum fitness. Should there be two mountain peaks rather close to one another, then a large change in the 'organism' could move it into the vicinity of the other peak and, although the target area is no doubt small compared with the vast field of possibility, there is a small chance that the fitness of the new position is greater than that of the old. It might, for example, be that the other peak is higher and more massive. If this was the case, then larger mutations might, from time to time, play an important part in evolution, although only a minority of such mutations would be expected to do so.

This counter-argument is not defeated by appeal to the piece of maladjusted machinery. That analogy works if the machine has, as most do, only one optimal adjustment. However, a microscope, for example, might be trained on a slide in which several different focal planes were all equally interesting. In that event, arbitrarily moving the coarse focus does have a finite, although not large, chance of bringing something interesting into better focus than has been attained so far. Moving the fine focus has, of course, no less a chance of improving the focus in the immediate present vicinity, and by analogy the effective occurrence of large mutations in no way rules out the kind of gradual evolution that Fisher envisaged.

Under what circumstances will organisms have several different optimal adjustments? They will do if genes interact non-additively in their effects on fitness, and it is on this subject of gene interaction that Fisher's argument in the *Genetical Theory* ceases to be self-consistent. In deriving the Fundamental Theorem, Fisher made the reductionist assumption that the effects of all genes on the variance could simply be added, which is, in effect, the assumption that there is no interaction between genes, not even the interaction between alleles which we call 'dominance'. The kind of interaction that produces multiple adjacent peaks of fitness is one in which the genotypes *AABB* and *aabb* are the most fit, and *AAbb* and *aaBB* are the least fit. Fisher did admit this kind of interaction to his argument, showing that it would result in tighter linkage between genes, and hence (as recombination still occurs) arguing that sex and recombination were maintained by selection in favour of speeding up the flux of evolutionary change (loss of recombination tends to put the brakes on – Maynard Smith 1978), but he never brought this part of his argument into confrontation with his Fundamental Theorem or with his case against large mutations. To have fully admitted a multipeak system into his theory, would have forced Fisher into conceding rather a lot of ground to Mutationism, and would also have admitted the first premise of Wright's Shifting Balance Theory. The one kind of gene interaction that Fisher did admit into his general argument is one that does not produce multiple peaks; it is now known as *specific modification*.

'Not the wing of a butterfly'

The one thing that can stand decisively against a scientific theory, be it the most ingenious, is empirical evidence to the contrary. Mimicry had come to provide the best empirical evidence in favour of Mutationism. The affair had started when E. B. Poulton (1908), stung into some very impolite polemics with Bateson, challenged him with one of Darwinism's best examples: the protective, mimetic resemblance of a defenceless insect to a noxious one. How, Poulton had asked, could an essentially random process like mutation, in effect paint on the wings of one butterfly the pattern of another?

The challenge must have seemed unanswerable, but answer it Punnett did, with considerable success, first in *Mendelism* (third edition, 1911), and then at greater length in *Mimicry in Butterflies* (1915) (hereafter referred to as *Mimicry*). Punnett set out to investigate the evolution of mimicry by a good Mendelian technique: crossing different mimics, and finding out how the difference was inherited. Ideally, one would want to cross different species, which was, and largely remains, impractical. Even the crossing of races of the same species in those days before commercial air-travel, must have presented insuperable problems. Punnett was unable to back his belief (*Mendelism*, third edition, p. 34) that differences between different species of model were the result of single mutations. He turned instead to mimics that were polymorphic, and carried out his own field-work in Ceylon (Sri Lanka) on *Papilio polytes*, a mimetic swallowtail with three forms of female, one non-mimetic (and resembling the male), the other two copying two probably distasteful 'poison-eater' swallowtails, *Pachliopta hector* and *Pachliopta aristolochiae* (at that time placed misleadingly in the same genus as their mimic).

Within a polymorphic species it was possible to perform genetic experiments, but the polymorphism itself was not a universal blessing for Punnett. The new calculations which he had commissioned from Norton showed how rapidly natural selection would substitute a more advantageous form for a less advantageous one and Punnett could not understand why all three forms remained in the population. He was forced, rather reluctantly, to the conclusion that the mimicry no longer worked – his field observations indicated that the mimics and models were distinguishable in flight and had different altitudinal ranges – and that the three forms were now selectively neutral. It was the best conclusion he could draw in the current state of knowledge (the better solution, that the polymorphism is balanced by negatively frequency-dependent selection would not appear until Fisher's mimicry paper of 1927), but has unfortunately led some later workers to think that Punnett believed that all mimicry was produced by mutational accident and had nothing whatever to do with natural selection:

> Punnett . . . knowing that . . . the forms differed by single allelomorphs . . . concluded that the mimicry did not evolve gradually and did not confer any advantage or disadvantage to the individual (Clarke & Sheppard 1960).

This is unhistorical and unfair.

Genetic experiments on *P. polytes* carried out by Fryer (1913) showed that the three forms were each produced by a single Mendelian gene [Fryer's interpretation of the inheritance pattern was much later shown by Clarke & Sheppard (1972) to be not entirely correct, but this in no way alters the main point]. The conclusion that each had been produced by a single large mutation seemed unavoidable.

> . . . if we take this view, which is . . . consonant with the evidence before us, we . . . cannot suppose that natural selection has played any part in the *formation* of a mimetic likeness. The likeness turned up as a sport quite independently of natural selection. But . . . natural selection . . . may nevertheless have come into play in connection with the *conservation* of the new form. If the new form possesses some advantage over the pre-existing one . . . is it not conceivable that natural selection will come into operation to render it the predominant form? (*Mimicry*, p. 92).

Similar results had been obtained for three forms of the polymorphic *Papilio memnon* in Java and for various forms of the African *Hypolimnas dubia*; again, all forms segregated clearly as single Mendelian genes and no intermediates appeared when they were crossed.

Fisher's theory of minimal genetic change therefore faced a serious empirical challenge unless he could show that it could account for these findings on the genetics of mimetic butterflies. Most readers will know that his solution was his famous theory of the 'modifiability of gene action', which became the pivot of his theory of the evolution of dominance (Fisher 1928). According to Fisher, although the fact is not obvious from the published papers of 1927 and 1928, it was solving the mimicry problem which first gave him this idea.

> With respect to my own work, it might be worth while referring to the paper of 1927, 'On some objections to mimicry theory: statistical and genetic' . . . where the notion of a gene acting as a switch was first developed . . . I should not like people to think that my interest in the modifiability of gene action was confined to, or dated from, the 1928 paper on Dominance. It would be truer to say that in 1928 it first occurred to me that *even* in respect of dominance the effect of a factor [gene] was conditioned by other factors (Fisher to K. Mather, 1942, *Letters*, p. 238).

The basic idea that natural selection might modify the system of inheritance itself had in fact been proposed by Wallace, interestingly enough in the very context of butterfly mimicry. He suggested that patterns originally inherited by both sexes might, through selection, become limited to females only (see Kottler 1980). The young R. A. Fisher (1920) had even toyed with the notion that such second-order effects of selection might have generated the apparent Mendelian genes from some more fluid or plastic hereditary substance.

In the *Genetical Theory*, and in the paper of 1927, on which its mimicry chapter is based, Fisher cites W. G. Castle's classic experiments with hooded rats. This piebald pattern of brown and white patches is produced by a single gene, but by selective breeding Castle was able to produce considerable changes in the hooded pattern. The crucial finding was that the hooded gene itself had remained unaltered: the changes in the pattern were produced by alterations in the rest of the genome.

The second part of Fisher's solution came from butterflies which were polymorphic for a rather simple pattern difference produced, as far as any one could tell, by a single gene. The orange and white forms of the females of certain sulfur butterflies (*Colias*) had been particularly well studied. The change of colour from orange to white could be thought of as a 'large' change and was no doubt a larger change at a single step than many a Victorian gradualist would have imagined, but Fisher was in the business of incorporating the findings of genetics into evolutionary theory, not of denying the results of genetic observation. Changes of this kind were obviously taking place in evolution and they were quite simple changes of pigment. What Fisher wished to dispute was that the whole of a mimetic pattern would appear in this way. However, polymorphisms could be maintained not just by the waning of the selective advantage of a mimetic form as it became commoner, but as Fisher and Haldane had both shown, simply by the heterozygote happening to be fitter than both the homozygotes. Such a polymorphism could therefore persist in a population without the patterns having any mimetic function whatever. Now, suppose that the ancestral *Papilio polytes* had two such forms in its population. There was, said Fisher,

> . . . no reason whatever on genetic grounds to believe that the combination [of genes which now produces the *aristolochiae*-mimic] on its first appearance at all closely resembled the modern [mimetic] form . . ., or was an effective mimic of *P. aristolochiae*; nor that the combination [now producing the mimic of *P. hector*] resembled the modern form . . ., or was an effective mimic of *P. hector*. The gradual evolution of such mimetic resemblances is just what we should expect if the modifying factors, which always seem to be available in abundance, were subjected to the selection of birds or other predators (*Genetical Theory*, p. 166 [p. 185]).

Both forms, in short, were originally non-mimetic. Suppose now that one of the forms had some slight resemblance to *P. aristolochiae* and underwent gradual evolution in the direction of the model's pattern. The other form, having no resemblance at all to the model, would remain unaltered. Now comes Fisher's own particularly original insight. Suppose that the genes which produce the increasing resemblance to the model were specific in their effect to just the one form of *polytes*; that is to say, they altered its pattern toward mimicry of the model, but had no effect whatever on the pattern of the other form. The unaltered form would finish up being the present-day male-like female, unchanged since before *polytes* became mimetic; the other form, after a long period of gradual modification, would be rather a good mimic of *P. aristolochiae*. By a similar process, either happening simultaneously or later, the form resembling *P. hector* would also have been produced from a third, originally non-mimetic form. (I have somewhat expanded Fisher's argument, in the hope of making it clearer. He has simplified the argument by ignoring the non-mimic, which I find makes the theory harder to explain.)

The crucial point is that the genes which produce the improvement in the mimetic resemblance must have their effects on that form only – Fisher called them 'modifying factors'; they have since come to be called *specific modifiers*. They have an important side-property from Fisher's viewpoint in not creating multiple peaks of the fitness surface. Fisher considers how a polymorphic species can improve each of its forms independently of the others. It was not too difficult to imagine that a species with only one form could modify that in the direction of better mimicry; Fisher had proposed a mechanism by which any number of forms, all coexisting in the same population, could each be modified into increasingly accurate mimics of several different models.

This theory therefore explains the results of the genetic experiments with polymorphic mimics. Just as in the experiment with hooded rats, the genome of the species has been altered, but the original gene (the *hooded* allele in the rats, the gene producing the original colour polymorphism in the butterfly) has remained, and is itself unchanged. It is like tuning the selector buttons (or dial) on a television set. Each button, or setting on the dial, has to have an independently tuned device behind it, which will lock the set onto a particular television station. Before the tuning is carried out, switching from one channel to another will produce nothing but snow on the screen. After tuning, one can get a perfect picture from each channel. A child who knew nothing about television sets would conclude that the button itself had been manufactured to achieve perfect reception: in fact, it is merely acting as a simple and mundane switch between two carefully tuned circuits. Likewise, after the specific modifiers which improve the mimicry have been built up in the population, the original gene which once merely changed the colour of the butterfly in some minor way, will now act as a switch, which will determine whether the butterfly produces one or other mimetic pattern (or indeed, the ancestral non-mimetic pattern). In genetic experiments, the patterns will segregate under the control of a single gene.

As Fisher summarized his argument:

> . . . it is the function of a Mendelian factor [i.e., a gene] to decide between two (or more) alternatives, but . . . these alternatives may each be modified in the course of evolutionary development, so that the morphological contrast determined by the factor at a late stage [in evolution] may be quite unlike that which it determined at its first appearance. The inference . . . that because a single factor determines the difference between a mimetic and a male-like form in *P. polytes*, therefore the mimetic form arose fully developed by a single mutation, is one that cannot fairly be drawn . . . (*Genetical Theory*, pp. 164–5 [pp. 183–4]).

In other words, the fact that there is a single gene difference between two forms when they are crossed cannot of itself take us to the conclusion that the forms have been produced by that mutation alone. Indeed, the major mutation that one detected by genetic experiments in a polymorphic mimic had originally had a much smaller effect on the pattern, now much exaggerated by the modifiers, and had originally nothing whatever to do with the mimicry: it was simply the residue of a polymorphism which was originally present for totally different reasons.

However, there was one further severe challenge thrown out by Punnett which Fisher had to meet: a particularly telling version of the old and familiar problem of the utility of incipient structures, or the 'unbridgable gap'. It was difficult to imagine what possible use part of a mimetic pattern, or a merely vague resemblance to the model would be. It was known that mimics frequently differed considerably in pattern from their camouflaged relatives and therefore appeared to have undergone considerable changes to reach the mimetic pattern; but the early stages would have been impossible to achieve under natural selection.

Darwin had been aware of this problem and had suggested that mimicry had commenced at some remote time in the past when both model and mimic were cryptically coloured. As the model gradually developed its warning colour, so too did the mimic, eventually being pulled, but entirely gradually, into the possession of a full warning pattern markedly different from that of its still camouflaged and conservative relatives. The apparently unbridgable gap was the outcome of this process, not a hiatus that the mimic had somehow jumped across (if I am on one bank of the river and my sister on the other, it does not follow that one of us swam over: we may have walked down opposite banks from the source).

Punnett found a system in which this theory could only work by the most disagreeable mental contortions: a mimicry ring. It occurred in Sri Lanka (Ceylon) and consisted of two Danaine models, which appeared to be rather inaccurate Müllerian mimics of each other, and three apparently Batesian mimics (Fig. 1), *Argynnis hyperbius*, *Elymnias hypermnestra*, and *Hypolimnas misippus*. The catch for the Darwinian theory was that the males of the Batesian mimics were not mimetic – a not uncommon condition – and could fairly be taken as showing the ancestral pattern of the mimetic species. In Darwin's scheme, one could suppose that all the species, both models and mimics, started with the pattern of the male *Argynnis*, or with the pattern of the male *Hypolimnas*, or with the pattern of the male *Elymnias*, but hardly that they started with all three patterns at once! This argued strongly that there had been a gap to cross, and that it must have been achieved by a single mutational step.

'One, two and three in your bosom'

Fisher, with his usual and unusual percipience, saw that the key to the whole problem was the evolution not of the mimics, but of the two models. Provided these could have become Müllerian mimics of one another by slow and imperceptible steps, then the whole ring could have evolved by the Darwinian scenario. Each model would have started with a different cryptic pattern, each somewhat resembling the cryptic pattern of one of the mimics. As the warning colour of the models gradually appeared, so did the mimetic patterns of the mimics; at the same time the two models gradually converged to become Müllerian mimics, dragging their Batesian mimics along with them to form the completed mimicry ring. In general the scheme will work provided only that the ring does not contain more, taxonomically diverse, mimics than it contains models. Punnett could have protested that three into two wouldn't go: his ring had three mimics. Fisher was equal to this, by arguing that *Hypolimnas misippus* was suspected by

Fig. 1. The Sinhalese mimicry ring that Punnett used to argue against Darwin's gradual theory of evolution. Fisher showed that it could after all, have arisen gradually. The females of the three mimics (top row, right halves) (*Argynnis hyperbius*, *Hypolimnas misippus*, *Elymnias hypermnestra* (=*undularis*)), are good copies of the Müllerian mimicry ring (bottom row) that consists of *Danaus chrysippus* and *Danaus genutia* (=*plexippus*), but each mimic seems to have evolved from the markedly different pattern still exhibited by its own male (top row, left halves). (Names in brackets are the now obsolete names used by Punnett.)

an independent authority (E. B. Poulton, no less) of being distasteful in its own right, and therefore legitimately left out of the theoretical argument.

The centre of debate now moves, therefore, from the origins of Batesian mimicry to the origins of Müllerian mimicry. The currents to be followed are more complex, for the former traces its history back in a fairly simple way to the Darwinist–mutationist debate, while the main stream of Müllerian mimicry has an important headwater in what we would now see as a less than olympian contest between two Darwinists, both of them gradualists and both bug-hunters in the English tradition: F. A. Dixey and G. A. K. Marshall.

This lesser dispute arose out of that most classical of evolutionary pursuits: the construction of phylogenetic relations. Dixey (1896) had attempted to deduce these in Pierid butterflies by considering the way their colour patterns had been evolving, using as his main consideration the way certain elements of the pattern might have appeared or been enhanced to produce Müllerian mimicry with members of other groups. Dixey had assumed that the evolution of Müllerian mimics consisted of mutual convergence, so that the species would develop elements of each others' patterns. This he called 'Reciprocal Mimicry'. Marshall's observations on the patterns of African butterflies led him to doubt this principle and to believe that only one of the species of any mimetic pair had altered in the course of evolution, the other retaining its ancestral pattern.

Both arguments are presented with considerable skill, an excellent balance between theoretical argument and evidence from the real world, and not a little wit. '. . . Mr. Marshall has appealed to arithmetic, and to arithmetic he shall go' and 'Mr. Marshall has now shot his bolt. It has failed . . .'. The very proper consideration of particular mimetic butterflies which takes up the major part of the papers I shall not discuss, as constituting too thorough a diversion from the logical debate itself, which turns on the degree of protection afforded to an evolving mimetic pattern (Dixey 1896, 1909; Marshall 1908).

Neither Dixey nor Marshall had any very clear impression about the nature of variation and, naturally at this time, were unable to distinguish the concept of a new mutation from pre-existing genetic variation. Both accepted 'the assumption that mimicry has been built up by a gradual process of selection from comparatively small individual variations' (Marshall 1908, p. 101). However, the extent to which biologists had already appreciated the importance of Mendel's work for Darwinian theory, and how early an intellectual base had been laid down for what biological tradition has regarded as the much later Modern Synthesis can be seen from two comments of Dixey's. First, with 'If . . . a well adapted type must have arisen, not by one or more large mutations, but by a series of mutations both numerous and minute, we should wish to know how such mutations are to be distinguished from continuous variations' (Dixey 1907, quoted by Poulton 1908), Dixey shows himself to understand very well, long in advance of the Modern Synthesis or the mathematical treatments of Fisher, the attribution of continuous variation to Mendelian genes; what, because quality becomes quantity, I have called the Backward Hegelian Principle.

The point indeed seems to have been generally and very quickly understood. Punnett, as early as 1905 (*Mendelism*, p. 51) comments, 'Doubtless some of the so-called fluctuations [continuous environmental variations] are in reality small mutations'. The significance of the non-blending of genes for the Darwinian theory, a notion for which Fisher later claimed priority (see letter quoted on p. 165), was also known to Dixey. In 1909 he commented that in order to persist a new variation would not need 'reinforcement from its original stock', provided it was Mendelian.

Marshall (1908) set out to prove the universality of what he called 'the Batesian principle', by which he meant the convergence of one species onto the pattern of another, unaltered species, and the non-existence of Reciprocal Mimicry or 'the Müllerian principle'. His argument observes a very proper concern for the principle that, in order to argue for the adaptive evolution of a character, we have to show that not only is its final state beneficial to the existence of the organism, but that each stage in its development should be more beneficial than the one which went before. Marshall agreed that when two distasteful species had the same pattern, then both benefitted from the resemblance, but pointed out that an adaptive end-point did not guarantee an evolutionary pathway. If we supposed that there were two species with warning colours, alike in everything but abundance, and in every generation a constant number of any warning pattern had to be killed to educate the predators, then both species would lose the same number. If they shared the same pattern then that number would be shared between them, with adaptive advantages to both, but if we considered the pathway leading to this end, then it must consist of the less abundant species (B) approaching the pattern of the more abundant (A). For if A had a strength of 100 000 and B of 5000, and 1000 individuals were required to educate the predators, B would be losing 20 per cent of its numbers and A only 1 per cent. Any variety of B which resembled A would clearly be at an advantage to the original pattern of B, but this was not reciprocally true of variations of A that looked like B. '[Suppose] that 10,000 specimens of A simultaneously present a sudden marked variation in the direction of B . . ., what will be the effect of the Müllerian factor in this remarkable variety?' With 90 000 of the A pattern now, 15 000 of the B pattern, and 1000 of each still being destroyed, 'a simple calculation shows that the percentage of loss . . . will be six times greater in the variety of A than in the typical form of A. . . . The Müllerian factor is capable of converting B into a mimic of A, but it cannot cause A to mimic B'.

What if A and B were equally abundant? Marshall seems to have had difficulty in finding a direct arithmetical argument to cope with this point (if we again assume that 1000 individuals have the new pattern, then either A or B can converge on the other), but showed that the advantage accruing to the mimetic form of the rarer species decreased as the discrepancy in numbers was reduced, until in the limit there was no advantage to it when the numbers were equal. As before, the argument from the final state benefiting both species provided no proof of a workable evolutionary pathway to the final state.

If A and B are equal . . . any small variation from one towards the other will not practically affect the numerical relationships of the two species and will therefore have no mimetic value. . . . The essential condition for the origin of Müllerian mimicry is lacking (Marshall 1908).

Marshall's general conclusion then was that while Müllerian mimicry was real enough, it did not evolve as Dixey and Müller before him had supposed, by mutual, gradual convergence. Rather, the less abundant of the species always converged, still gradually, towards the commoner and if in the limit the two species were equal in abundance, then neither had 'an iota of advantage over the other' and mimicry would not evolve.

Dixey (1909) found a number of serious flaws in this argument. First, Marshall assumed that both species were equally unpalatable; if the rarer one was the nastier it could be the better protected. This argument is something of a smoke screen, as it makes no difference to the main point if, as Dixey concurs, we forget about abundance and consider only the percentage lost in each generation. Second, Marshall had left open the possibility of mutual convergence by admitting that the relative abundance of the species might fluctuate, so that first A approached B and then B converged on A. This did not alter the theoretical point, but was more material to the interpretation of phylogeny which was Dixey's interest.

The third point is the critical one. Marshall has assumed that the new form of A gains some protection from resembling B, and loses all its protection from resembling A. But if we are dealing with small continuous changes, this is wrong. The new form will still somewhat resemble A. Let us suppose two equally common species, each with two forms: A is non-mimetic, but its form Ab somewhat resembles B. Similarly, B's form Ba looks rather like A, and of course, rather like its main form B, which the birds do not confuse with A. Classes Ab and Ba will be protected from birds that have encountered either of them, or A, or B, whereas A and B are protected only against birds that have encountered their own class. The mutually mimetic forms, Ab and Ba, are clearly at an advantage in both species. Only if the ancestral form vanished on the appearance of the mimetic variety (and 'immeasurable instances of the persistence of an ancestral form are known throughout organic nature', says Dixey), would Marshall's argument be right.

The victory goes to Dixey if we accept, as Marshall seems to, the premise that all variation is small and continuous. Marshall's argument is valid if the new variety of A, the more abundant species, is a saltation of such magnitude that, while mistaken for the different pattern of B, it is no longer confused with the wild-type of A. It is tempting to believe that Marshall was in a small way a 'new man' who had perceived this important implication from the science of genetics, but there is no internal evidence of this in his paper.

Marshall's argument then is correct, if we assume that the new variety is produced by a major mutation. Punnett stood this assumption on its head in *Mimicry* and argued that Marshall's arithmetic proof about the relative advantage to species A and B actually proved that a major mutation must occur during the evolution of Müllerian mimicry.

Though it is difficult to regard Batesian mimicry as produced by the accumulation of small variations through natural selection, it is perhaps rather more plausible to suppose that such a process may happen in . . . Müllerian mimicry. For since the end result is theoretically to the advantage of both species . . . variations on the part of each in the direction of the other would be favourably selected . . .

Difficulties, however, begin to arise when we enquire into the way in which . . . their meeting one another half-way . . . may be conceived of as having come about. By no one have these difficulties been more forcibly presented than by Marshall . . .

[there follows a thorough account of Marshall's arithmetical argument]

. . . the Müllerian factor . . . cannot bring about a resemblance . . . of a more numerous to a less numerous species. Further . . . there can be no approach of one species to the other when the numbers are approximately equal. A condition essential for the establishing of a [Müllerian] resemblance, no less than [a Batesian resemblance], is that the less numerous species should take on the pattern of the more numerous. Consequently the argument . . . against the establishing of such a likeness by a long series of slight variations is equally valid for Müllerian mimicry (*Mimicry*, pp. 72–4).

In turning Marshall's argument round on itself, Punnett had produced an argument that was circular, in the strict sense of incorporating the conclusion in the premises. Punnett surely did not perceive this, but was rather so convinced of the role of large mutations that the rightness of a saltational reading of Marshall was to him self-evident.

This apparent proof of saltation for Müllerian mimicry, Fisher pointed out, was not an aside as Punnett had presented it, but the pivotal point of the whole argument. Fisher's task was therefore to prove Müller and Dixey right about the mutual gradual convergence of Müllerian mimics. He adduced no strong argument against Marshall, and only got as far as showing that gradual, mutual convergence between two Müllerian mimics was possible. He plays strictly by the rules of deduction and having proved the statement 'Dixey's theory is not necessarily untrue' does not explicitly state as a corollary 'Marshall's theory is invalid', although he surely believed it was.

Fisher's argument should have depended on equating mutual convergence with gradual evolution and one-way convergence as proposed by Marshall with saltation. This distinction he establishes by pointing out that one-way convergence requires the new form of the species that does the converging, to lose all the advantages of being mistaken for a member of its own species. It must therefore have undergone a rather large change. He then establishes the important concept of a normal curve of error on the part of the predators; that is to say, discrimination is not an all or none phenomenon, but that given the ambient variation in the opportunity that a bird has for observing a mimic, its tendency to attack it will vary from time to time, with the patterns closest to the model escaping attack most often. The editors have suggested to me that Fisher here anticipates 'signal detection theory' (see also Duncan & Sheppard 1965), as he anticipated so many other things.

He goes on to consider the convergence of species A toward species B and *vice versa*. Marshall would maintain that mutants of B, the less protected species, which resembled A would be at an advantage, but not the reciprocal mutants of A. We must, in considering whether A can converge toward B, Fisher says, consider not just mutations of A in the direction of B, but mutations of A in the diametrically opposite direction. Then as both classes of mutants lose equally by departing from the A pattern, but only the first gain from resembling B, these mutants will be favoured and species A will converge toward B. The argument, as worded, looks fallacious. There is no reason whatever to consider the fate of an equal and opposite mutation when modelling the fate of a new mutation. One must compare the fitness of the mutation against the original wild-type and this is always to the disadvantage of mutants of A in the direction of B, but not for mutants of B in the A direction, provided they are large enough. Marshall's argument for one-way convergence, following a saltation, stands. But what Fisher was considering was not new large mutations at all. He was concerned with genes within a multigenic system controlling continuous variation in colour pattern. Here, with additive gene action assumed for simplicity, the mean pattern of a '+' allele (already at high frequency) is roughly as far above the population mean as the '−' allele's pattern is below it. In this case, the '+' alleles of species B are favoured over the '−' alleles, and the '−' alleles over the '+' alleles in species A. Gradual mutual convergence of the two species will then take place. Thus if we use Fisher's Backward Hegelian Principle that continuous, quantitative variation has a Mendelian basis, we can indeed see that Dixey's model of mutual convergence is correct (the less-protected species B will in fact converge the faster and farther).

Thus Fisher, although his conclusion incorporates its own premises in the way that Punnett's did – this time by assuming from the start that he was dealing only with genes of individually very small effect – is on stronger ground. Punnett was trying to show that major mutations did produce mimicry. Fisher was not trying to prove conclusively that they did not, but only to show that mimicry provided no decisive evidence against his overall gradualist thesis: and for this he needed only to show that Müllerian mimicry could evolve gradually. A stronger proof was not needed.

However, the cautious way Fisher has worded the argument, refusing to state anything more than was logically necessary, has unhappily left the way open for anyone reading this section too quickly, or superficially – that is without the most painstaking analysis – to imagine that they have encountered one or other of two different mutually contradictory arguments: (a) that Fisher has disproved both Marshall's argument and the argument that major mutations were involved, or (b) that Fisher has argued, decisively for Batesian mimicry and with an open mind for Müllerian mimicry, that major mutations could be involved. Both these readings have been made and both are incorrect.

The *Genetical Theory*, as read today by most population geneticists is divided into three parts: the serious stuff at the beginning, the final chapters on eugenics (the less read the better – if we had been alive then we would all have known how misguided it was), and in between an excursion into a little by way of English butterfly-collecting, the chapter on mimicry. The eugenics chapters are, as Norton (1983) has shown, an integral part of the work, certainly producing an application of Fisher's pivotal gene-flux model of evolution, quite possibly giving Fisher the starting-point for a number of his ideas, certainly allowing him to entertain Hope. I hope that I have likewise convinced the reader that the chapter on mimicry was similarly no side track. It constituted an essential piece of Fisher's thesis: in showing that mimicry could evolve gradually after all, it removed a considerable, threatening, empirical, and theoretical obstacle to Fisher's whole gradualistic model. Failure to deal with the problem would have left Fisher with large arbitrary changes through mutation, a loss of the creative supremacy of natural selection and, what is more, a clearly 'multipeak' view of adaptation, which would open the way for the other ascendant theory of shifting balance evolution being developed by Wright.

Fisher dealt with the problem very skilfully. He also used his solution to develop a further, even more comprehensive view of selection, which allowed the process to become infolded, and creation to occur at an even higher level of complexity. The concept of modification, Fisher saw, could be applied to all genes. His best known application is his highly controversial theory of the evolution of dominance (Fisher 1928), but Fisher hoped for even wider applications. Specific modifications of the effects of beneficial mutations would lead as a general principle to their becoming more beneficial (*Genetical Theory*, p. 95 [pp. 102–3]). In this way natural selection did not simply create the organism, but created the properties of the evolving genetic system itself. Dominance was but one example; Fisher's tenacity in arguing for the evolution of dominance, which he bequeathed to his colleages E. B. Ford and P. M. Sheppard, and which may have been one of the crucial factors in his falling out with Wright (Provine 1985), came from his perception of this theory as the first step in the acceptance of a theory of selection as all-powerful not just at the first level – the organism – but at the second level of the genetic system.

But a saga has three parts: having told you of the origins of hero and the dragon, and how the one slew the other, I must now finish by saying what became of the dragon-gold when the hero brought it into the world of men.

'The wise world'

Consider the following evolutionary sequence, watching for a change in the way mimicry is explained.

We have no reason to assume that when genes such as these first appeared their effects were similar to those which they produce to-day. On the contrary, we may suppose that in a given palatable species a gene arose which chanced to give some slight resemblance to a protected form. This would gradually be improved by selection of the gene-complex, and the consequent alteration of the effects produced by all the genes acting together, until an accurate mimicry had been attained. Such a process would be one of slow continuous change, but at the end the profound difference so produced would still be under the control of a single factor; yet this would not mean that the mimic had arisen from the non-mimetic form suddenly by a single act of mutation (Ford 1931, p. 73).

Fisher . . . pointed out that the effects of a gene may be modified by selection operating on the total gene complex. Consequently, when any mutation chances to give a remote resemblance to a more protected species, from which some advantage, however slight, may accrue, the deception will constantly be improved by selection. This, working upon genotypic variability, will result in a gradual change in the *effects* of the gene concerned, which may come to be very different from those which it had at first. But the gene itself is unchanged, and remains as a switch turning on one or another set of characters subject to genotypic variability and, consequently, susceptible of selection. Thus, though the forms of a species may be controlled by a single factor difference, which must have originated suddenly by mutation, we are none the less entitled to regard them as the product of slow evolution (Ford 1937).

. . . the *effects* of genes chancing to give some slight superficial resemblance of mimic to model have been modified by selection operating on the gene complex: such genes themselves remaining unaltered, to act as a switch in maintaining the alternative forms (Ford 1937).

[Fisher's] solution . . . depends upon the fact that the *effects* of a switch-gene (as of other genes) are subject to genetic variation. They can therefore be modified by selection operating upon segregation taking place within the gene-complex, leaving the main gene itself unchanged. He therefore suggested that when a mutation chances to give an unprotected species some slight resemblance to a protected one, *the mutant gene would spread owing to the advantage it confers* while its effects would be gradually modified, and the mimicry for which it is responsible progressively improved, by selection acting on the gene-complex. If polymorphism be evolved, the original mutant would then remain as the switch-gene controlling alternative forms. Though it arose suddenly by mutation we are not to suppose that the adaptations for which it is responsible, in all their perfection, did so too; for these would have been attained by gradual evolution within the ambit of the major-gene (Ford 1964, pp. 234–5, second set of italics added).

. . . both Fisher . . . and Ford . . . take the view that when a mutant producing some mimetic resemblance is established in a population the resemblance is improved by selection for a gene-complex in which the original effect of the gene is altered towards more perfect mimicry (Clarke & Sheppard 1959).

Fisher and Ford . . . on the other hand have put forward a view intermediate between the extreme ones of Darwin on the one hand and Goldschmidt on the other. They [Fisher and Ford] maintain that a resemblance between the mimic and model as good as that often found in nature cannot reasonably be expected to arise by chance. They therefore suggest that a mutant becomes established which gives a sufficient resemblance to be advantageous and that this is gradually improved by the selection of modifiers which alter the effects produced by the mutant (Clarke & Sheppard 1960).

The reader, feeling like one subjected to interminable variations by allcomers on that theme of Paganini, is asked to note in this sequence of quotations that we can witness a genuine case of gradual evolution, not of the mimetic pattern, but of a theory about it. Starting more or less at Fisher's position, that all is gradual, we end with statements by Ford, and by Clarke and Sheppard which admit a quite large initial mutation into the process. The change is subtle, but important. For Fisher, the difference started as gene of slightly larger effect than the rest, existing at first as a polymorphism unconnected with the mimicry, whose effects were then modified and magnified. For Ford, Clarke, and Sheppard, as they finally stated the argument, the large mutation actually initiates the mimetic resemblance by jumping the gap between the old pattern and the new mimetic one. Operationally, the two theories are not easy to distinguish as, in a polymorphic mimic, they give the same outcome. They would, however, predict different results for monomorphic mimics, which in Fisher's scheme would have evolved entirely gradually (for there would be no initial large mutation), and would therefore differ polygenically from their closest relatives. In the later Oxford scheme, such mimics would normally be expected to contain a fairly major mutation which had initiated the mimicry.

It is difficult to tell just when the change of thought takes place. By the 1960s we are definitely with the second model, which Sheppard laid out explicitly and lucidly in a paper published in 1962 (but written several years earlier). It will be seen that Ford's earlier expositions are genuinely ambiguous: does the mutation that chances to give a slight mimetic resemblance initiate the mimicry or not? This tradition of ambivalent wording continues even with Clarke and Sheppard: it would be difficult to be certain which way to read the 1959 quote, were it not for the fact that the following passage occurs later in the same paper.

> The presence of [the non-mimetic form] *leighi* in some areas shows that a non-mimetic form of *P. dardanus* can be maintained in a polymorphic state. This raises the question of whether the evolution of a new mimetic form need wait for the appearance of a new mutant. It seems possible that a mimetic polymorphism could be evolved from a non-mimetic one if suitable models become available. For example, if a protected species *even remotely resembling leighi* should establish itself in South Africa, *leighi* might well evolve into a mimic of that species (Clarke and Sheppard 1959, italics added).

This is none other than Fisher's original theory, now being put forward as a new alternative to the main thesis of the paper! (At least one other author has since thought of this theory as a 'new' idea.) The Oxford School finally deviated so far from Fisher's line as to adopt multipeak portrayals of the fitness surface which, although they made no concessions to Wright in that they were used to represent changes under mutation and selection only without the participation of random genetic drift, were manifestly completely at variance with Fisher's single peak model and flux of tiny mutations (Clarke & Sheppard 1962).

I will try to answer two questions about this little piece of intellectual evolution: how it came about, and how the theory was always attributed to Fisher.

The theory, which I have called the two phase theory, was not new, even when Fisher was writing. It is recounted, clearly and at some length, in Punnett's *Mimicry*. There are, says Punnett, authors who take a position intermediate between Darwinism and saltationism.

> There are writers . . . who adopt a view more or less intermediate between those just discussed. They regard the resemblance as having arisen in the first place as a sport of some magnitude on the part of the mimic, rendering it sufficiently like the model to cause some confusion between the two. A rough-hewn resemblance is first brought about by a process of mutation. Natural selection is in this way given something to work on, and forthwith proceeds to polish up the resemblance until it becomes exceedingly close. Natural selection does not originate the likeness, but, as soon as a rough one has made its appearance, it comes into operation and works it up through intermediate stages into the finished portrait. It still plays some part in the formation of a mimetic resemblance though its rôle is now restricted to the putting on of the finishing touches (*Mimicry*, p. 71).

Who, one may ask, are these 'writers', and why must they remain anonymous? The theory was E. B. Poulton's. He had proposed it, using a particular instance rather than stating a general principle, in the middle of a refutation of Bergson, published in his own journal, *Bedrock*.

> If it be unreasonable to suppose that all these mimetic features arose spontaneously and together, what is the probable explanation of their origin? It is probable that by a spontaneous variation a white band [on the hindwing] appeared in the ancestral form . . . and that this was from the very first sufficient to confer some advantage by suggesting the appearance of a dominant Model . . . From this point Natural Selection acting on further variations produced the detailed likeness which we see in the white band itself and in the other mimetic features (Poulton 1912).

Poulton's solution was therefore that the gap could be crossed by a single mutation, which would not however have produced very accurate mimicry – just good enough mimicry to 'suggest' the pattern of the model to the predators. After that, gradual evolution, with natural selection acting on much smaller genetic variations, would bring the mimicry up to a high degree of perfection. It is not surprising that both the argument and its authorship were forgotten; Punnett's account was the best publicity it received, and although his refutation of the argument is only weak, he effectively buried Poulton in anonymity, out of, I suspect, a desire to be polite and to refrain from anything like personal criticism.

The two phase theory was again put forward by Nicholson (1927), in a paper whose great length has ensured it a very small readership. Were Fisher and the Oxford School aware of Poulton's and Nicholson's suggestions? They never cited Poulton; Nicholson, as far as I know, was cited only once by Sheppard, and then for a different reason. If they were aware of Nicholson's paper, they probably thought of him as supporting Punnett, for Nicholson had accepted that when butterfly mimicked butterfly, rather than when fly mimicked wasp, there would be close enough genetic homology between model and mimic for the initial mimicry to need no further improvement once the major mutation was established.

The question of Poulton's influence is harder to unravel. Ford was associated with him in Oxford at least to the extent of analysing Poulton's accumulated broods of the Mocker Swallowtail (*Papilio dardanus*) (Ford 1936), and Fisher both in the *Genetical Theory* and in letters acknowledges 'my esteemed friend the late Professor Poulton of Oxford . . . that very kind old man . . . who helped me much with the Mimicry chapter' (Fisher to L. P. Brower and E. B. Ford, *Letters* pp. 186, 196, 203). The insects used in the plates in the *Genetical Theory* came from the collections in Poulton's department at Oxford. It is hard to believe that Poulton's ideas on the evolution of mimicry were entirely without influence either on Fisher in Cambridge or on Ford in Oxford.

I can see three reasons why the Oxford School's opinion veered away for Fisher. First, there was the influence of J. B. S. Haldane, who had all along worked with a theory of evolution based on the action of single mutations of rather large effect, which is the foundation of the modern simplistic 'bean-bag' version of population genetics. The covert rivalry between Fisher and Haldane – who seldom cited each other's work – became diluted at second hand, and Sheppard, for one, respected Haldane's work and sought his help in the interpretation of difficult butterfly broods. Haldane had become interested in the possibility of powerful selective forces producing evolutionary stasis because they were balanced against each other (Haldane 1954), and had estimated that there must have been a selective differential of around 50 per cent to establish the gene producing the industrial melanism of the Peppered Moth in the time observed.

Second, there was the outcome of Ford's programme for studying the behaviour of genes in natural populations (Ford 1969). This revealed large, detectable selective differences not only in the Peppered Moth – close to the value predicted by Haldane – but in the snail *Cepaea* and the Scarlet Tiger Moth as well (Kettlewell 1955; Cain & Sheppard 1954; Fisher & Ford 1947). The Oxford School maintained, with some lack of logic – for the discovery that *some* genes were strongly selected could not prove that *all* genes were strongly selected – that this backed Fisher's thesis on the unimportance of random drift. Yet Fisher had assumed, as an integral part of his gradualist 'flux' model, that only genes of very small selective effect were taking part in evolution. It was migration and large overall population size that prevented the effective drifting of these weakly selected genes. The Oxford School had continued to agree with him on the absence of drift, but for a completely different reason: it was rendered ineffective by strong natural selection, which, as it would hardly do to maintain that the cases they had investigated were altogether exceptional, was assumed to be universal and all-pervasive. It would follow that a large proportion of the genes taking part in evolution were mutations of large selective value and perhaps of large phenotypic effect. Here was a difference with Fisher which the Oxford School played down by presenting Fisher's belief in small selective differences as simply the proper caution in view of the lack of evidence, in 1930, for strong selection; they chose on the whole not to notice how central the idea had been to Fisher's grand schemes.

These two factors, Haldane's theories and, particularly, their own empirical findings, led the Oxford School to think easily in terms of genes of rather large effect, producing large differences in fitness, and to incorporate them naturally into their thinking about mimicry.

Third, Goldschmidt's (1945) approach to the mimicry problem, although they were at pains to refute it, cannot have been without influence. He had argued, in line with his thesis that species arose by saltation, that mimetic patterns did also, and that they were fully mimetic *ab initio* on account of the limited number of routes which development could take. The Oxford School found good experimental evidence in their butterflies to refute Goldschmidt's conclusion that no further modification of the pattern would take place, by crossing mimetic with non-mimetic races, and showing that the mimicry broke down on a genetic background where the major mimicry gene was not found (Clarke & Sheppard 1960, 1963). Goldschmidt's argument that because rate-altering mutations could be shown to produce spectacular alterations in wing-pattern, it therefore followed that they could easily produce a particular, pre-determined pattern – as is required if the result is to be a high quality mimic – was none other than the old bogey, saltational change which is as miraculous as creation; this could be rejected rather easily on logical grounds. But the Oxford School were clearly not completely unresponsive to Goldschmidt's other suggestions, that when the species were closely related the mimicry might be produced by homologous developmental pathways or even homologous genes – although this aspect did not appear until they started to experiment with *Heliconius* (see Sheppard *et al.* 1985) – and, more fundamentally, that mutations of large effect could rather readily convert one pattern to another very different one. Their final position could fairly be described as a synthesis of the saltational views of Punnett and Goldschmidt on the one hand, and the extreme gradualism of Fisher on the other.

The drawback with that account of what had happened is of course that Poulton and Nicholson had made the same synthesis long before. The Oxford School could indeed take much credit for having worked the matter through experimentally, while Poulton and Nicholson had courted obscurity by doing nothing further with their ideas; but the oddity of the Oxford School's attributing the whole of the idea to Fisher still needs to be explained.

Even given that Poulton and Nicholson had been forgotten, it is surprising that the Oxford School in presenting their theory to themselves and to the world as Fisher's, had so thoroughly lost sight of the extreme gradualism that had been pivotal to the *Genetical Theory*. I made what I hope is the last mis-reading of Fisher when I ascribed (Turner 1977) the two-phase system jointly to Fisher and Nicholson. Partly, the reason lies in the style of writing scientific papers, in which a clean, simple line is favoured: set up two people and their theories, in this case Fisher and Goldschmidt, support one and refute the other. Partly it lies in a shift in the importance attached to different parts of Fisher's theory: it is the concept of modification, not the gradualism, that is held to have been crucial – and certainly it represented a considerable advance on the Poulton–Nicholson theory. Finally, it results from the cautious way Fisher words his argument, not nailing his colours to the absence of saltation in the way that Goldschmidt nailed his to the absence of modification. Fisher shows that evolution can be entirely described in his gradualistic terms, does not give any positive consideration to the consequences of a less gradualistic model, and advances a statistical argument against the role of large mutations. A reading of the overall message is clear enough, but a careful reading will, I think, reveal that nowhere does Fisher actually say that large mutations cannot be considered as part of evolution in general or of mimicry in particular. Whether this is a brilliant bit of scientific politics, or the mathematical mind refusing to assert more than can be proved in strict logic, it is this tension between intention and statement that makes the *Genetical Theory* so infuriatingly difficult to read. It certainly made it possible for the Oxford School to attribute their transformed Fisher theory to Fisher.

Lastly, of course, we may all simply have mis-read him!

'And so grow to a point'

It is not, I think, overstating the case to say that the Oxford Ecological Genetics School had built up a picture of evolution – we could call it the Oxford Model – which was empirically based, very much their own, and a synthesis of the views on the one hand of Punnett (or even perhaps Goldschmidt), and of Fisher and Poulton on the other. To a great extent they went about this by stealth. Their debt to Punnett was indirect and was never acknowledged. Ford (1957) had felt that Punnett had continued to promulgate a 'dangerous fallacy', and when a part of that argument turned out not to be fallacious, he could not be expected to see – who would? – that Punnett's views were partly correct. To have trumpeted the discovery of the importance of rather large mutations as a radical departure from Fisher's theory would have been an act of personal disloyalty to a friend and mentor. Rather, the fiction was maintained that Punnett was totally done for and Fisher vindicated. As we have seen, this stance was maintained by a delicate balancing act: by emphasizing the importance of specific modification as a seminal insight – which it was – and by writing non-committally and with great ambiguity about the size and function of the initial 'large' or 'slight' mutation.

The contribution which Poulton had made to the subject was also largely ignored. Their view of him was, not quite justly, that he did not understand genetics (Ford to Fisher, 1955, *Letters*, p. 203); Punnett, they felt, understood nothing else.

The Oxford School could then have claimed credit for a new synthesis, if not quite a 'direction' in the theory of evolution, but being Oxford gentlemen, did not. The kind of restraint considered the mark of good breeding left them unable to conduct the kind of publicity campaign, complete with its own title and buzzwords which, more now than then, has become necessary for any scientist who thinks he or she has a new idea, to be heard above the general hubbub. [There are of course, exceptions: a really important theory will surface of its own accord in the end, but the point I am making is backed by the length of time that it took for Hamilton's work on kin selection (Hamilton 1963) to catch the general imagination.]

Scientists, as much as politicians, interpret history to their own service. Setting up rival schools and confirming one while rejecting the other, not only conforms to the accepted norms of hypothesis testing: it is more fun, and a lot easier than non-whiggish historical analysis. It is fashionable to see a grand philosophical divide in evolutionary biology between gradualism, adaptationism, and liberal capitalism on the one side, and a less adaptationist, more Marxist saltationism or punctuationism on the other. The story goes that after an initial softening ('pluralism') toward the Marxist heresy around 1930, the mainstream of orthodox evolutionary theory 'hardened' back toward gradualism (Gould 1983).

Things were more complicated than that. It is easy enough to see a grand debate between saltationists and Darwinists at the beginning of the century, whose repercussions continued for a long time, with R. A. Fisher declaring firmly for the Darwinist camp. But it is important to see that something quite complex happened between 1900 and 1930: a debate which had tended to be about mutationism or saltationism at all levels from the gene to the species, divided into a series of questions about gradualism: was evolution gradual at the genetic level; were species evolved gradually; was fossil evolution gradual? Fisher, arch-gradualist in genetics, was willing to concede punctuated evolution in the longer term:

> . . . I should like to keep a mind open to the possibility that . . . the primary features [of bats] . . ., the development of the wings and the habit of preying on insects in flight, might have been developed quite rapidly, if we take an evolutionary scale of time . . . Osteologists . . . must often be stressing features which have arisen quite rapidly, and in a sense casually, in . . . being slightly useful to the parent species at a time when its food happened to have some peculiarity . . . From being so far a heretic on the fossil evidence, I am debarred from relying on it in support of what I certainly think may be true, i.e. the rapid origination of the primary distinctions of great classes (Fisher to L. Darwin, 1932, *Letters*, pp. 149–50).

At the genetic level, Fisher's stand, which was fairly extreme even at the time, was followed by what could be called 'the softening of the modern synthesis' or 'synthesis by stealth'. The Oxford School, while remaining firmly opposed to a creative role for random drift, and therefore thoroughly adaptationist in this sense, compromised with the mutationism of Goldschmidt and Punnett. They declared themselves 'for' Fisher and against the others, when they could have declared that all were equally right (or if wanting to be uncharitable, wrong), by emphasizing the points on which they believed Fisher was right, and glossing over or dismissing those of his views which they saw as incorrect. This again is a characteristic scientific approach to history.

There is a significant difference in the Oxford response to Fisher's two adversaries: with genetic drift it was 'all the way' with Sir Ronald (Turner 1985). When it came to the size of mutational changes, and the strength of selection which could operate on beneficial mutations (naturally, no one doubted that the deleterious ones were strongly selected) they surreptitiously, but decisively abandoned him, while maintaining both to themselves and to the world the fiction of loyalty. I suggest that this difference was dictated largely by the Oxford School's own experimental findings: the occurrence of strong and manifest natural selection operating on single genes of large phenotypic effect. As English empiricists and for the most part agnostics, if they thought about Fisher's grand philosophical schemes and his tough, but ultimately kind Creator, they said nothing on the subject in public.

The modern attempt to revive saltationism or punctuationism against gradualism, seems to me to have inadequately distinguished between the levels of biological organisation at which the debate can be conducted. In seeking to revive something resembling Goldschmidt's view of the species as mutant, its proponents have glossed over the important qualification made by the study of mimicry: that however 'large' a mutation may be which might carry an organism into a different adaptive zone, the adaptation is unlikely in the extreme to be perfect *ab initio*, and hence that evolution cannot be attributed solely (or if one likes, even chiefly) to this kind of mutation. Maynard Smith (1983) expresses what I believe is now a common view, that there is nothing inherently impossible in a 'monstrous' mutation taking part in evolution; although such events might be rare, in the vast expanse of evolutionary time there might have been many of them, and only a few might be needed to explain some of the very occasional 'happenings' of evolution, as when the land or air were first colonized. What most evolutionary geneticists would dispute is not the size of the mutational change, but its ability to produce a fully perfected adaptation *ab initio*.

It is here that we need the important evolutionary principle that was forged in the debate over mimicry, generated by Poulton and Nicholson, and considerably refined by Fisher: that evolution can operate on a compromise, with monstrous mutations being subsequently modified and improved.

This I suggest, constitutes the rather quietly achieved Synthesis of the Oxford School of Ecological Genetics.

Acknowledgements

Whatever merits this paper may have, it owes to long discussions over the last few years with Jonathan Hodge, William Kimler, and Robert Olby. In particular Hodge drew my attention to the importance of Fisher's Christian faith, an aspect of his thinking almost totally ignored by his fellow scientists because most of them found it totally unsympathetic. I am also most grateful to Mark Ridley for his constructive suggestions.

References

Allen, G. E. (1983). The several faces of Darwin: materialism in nineteenth and twentieth century evolutionary theory. In *Evolution from Molecules to Men* (ed. D. S. Bendall), pp. 81–102. Cambridge University Press, Cambridge.

Bennett, J. H. (ed.) (1983). *Natural Selection, Heredity, and Eugenics. Including Selected Correspondence of R. A. Fisher with Leonard Darwin and Others.* Clarendon Press, Oxford.

Bowler, P. J. (1983). *The Eclipse of Darwinism. Anti-Darwinian Evolution Theories in the Decades around 1900.* Johns Hopkins University Press, Baltimore and London.

—— (1984). *Evolution. The History of an Idea.* California University Press, Berkeley.

Brush, S. J. (1976). *The Kind of Motion we call Heat. A History of the Kinetic Theory of Gases in the 19th Century. Book 2. Statistical Physics and Irreversible Processes.* North Holland, Amsterdam.

Cain, A. J. and Sheppard, P. M. (1954). Natural selection in *Cepaea. Genetics,* **39**, 89–116.

Clarke, C. A. and Sheppard, P. M. (1959). The genetics of some mimetic forms of *Papilio dardanus,* Brown, and *Papilio glaucus,* Linn. *J. Genet.* **56**, 236–60.

—— and —— (1960). The evolution of mimicry in the butterfly *Papilio dardanus. Heredity,* **14**, 163–73.

—— and —— (1962). Disruptive selection and its effect on a metrical character in the butterfly *Papilio dardanus. Evolution,* **16**, 214–26.

—— and —— (1963). Interactions between major genes and polygenes in the determination of the mimetic patterns of *Papilio dardanus. Evolution,* **17**, 404–13.

—— and —— (1972). The genetics of the mimetic butterfly *Papilio polytes* L. *Philos. Trans. Roy. Soc. Lond. Ser. B,* **263**, 431–58.

Dawkins, R. (1983). Universal Darwinism. In *Evolution from Molecules to Men* (ed. D. S. Bendall), pp. 403–25. Cambridge University Press, Cambridge.

Dixey, F. A. (1896). On the relation of mimetic patterns to the original form. *Trans. Entomol. Soc. Lond.* **1896**, 65–79.

—— (1909). On Müllerian mimicry and diaposematism. *Trans. Entomol. Soc. Lond.* **1908**, 559–83.

Duncan, C. J. and Sheppard, P. M. (1965). Sensory discrimination and its role in the evolution of Batesian mimicry. *Behaviour,* **24**, 269–82.

DuNouy, L. (1947). *Human Destiny.* Longmans, Green, New York.

Fisher, R. A. (1911). Heredity. Unpublished paper, read to the Cambridge University Eugenics Society. (Printed in Bennett 1983, pp. 51–8.)

—— (1912). Evolution and society. Unpublished paper, read to the Cambridge University Eugenics Society. (Printed in Bennett 1983, pp. 58–62.)

—— (1918). The correlation between relatives on the supposition of Mendelian inheritance. *Trans. Roy. Soc. Edin.* **52**, 399–433.

—— (1920). 'Balanced lethal' factors and *Oenothera* 'mutations'. *Eugen. Rev.* **11**, 92–4.

—— (1927). On some objections to mimicry theory – statistical and genetic. *Trans. Roy. Entomol. Soc. Lond.* **75**, 269–78.

—— (1928). The possible modification of the response of the wild type to recurrent mutations. *Am. Nat.* **62**, 115–26.

—— (1930). *The Genetical Theory of Natural Selection.* Clarendon Press, Oxford.

—— (1934). Indeterminism and natural selection. *Philos. Sci.* **1**, 99–117.

—— (1947). The renaissance of Darwinism. *Listener* **37**, 1001 and 1009; *Parents' Rev.* **58**, 183–7.

—— (1950). *Creative Aspects of Natural Law*. The Eddington Memorial Lecture. Cambridge University Press, Cambridge.

—— (1954). Retrospect of the criticisms of the theory of natural selection. In *Evolution as a Process* (eds J. S. Huxley, A. C. Hardy, and E. B. Ford), pp. 84–98. Allen and Unwin, London.

—— (1958). *The Genetical Theory of Natural Selection*. (2nd edn) Dover Publications, New York.

—— and Ford, E. B. (1947). The spread of a gene in natural conditions in a colony of the moth *Panaxia dominula* L. *Heredity*, **1**, 143–74.

Fisher Box, J. (1978). *R. A. Fisher. The Life of a Scientist*. John Wiley and Sons, New York.

Ford, E. B. (1931). *Mendelism and Evolution*. Methuen, London.

—— (1936). The genetics of *Papilio dardanus*. Brown (Lep.). *Trans. Roy. Entomol. Soc. Lond.* **85**, 435–66.

—— (1937). Problems of heredity in the lepidoptera. *Biol. Rev.* **12**, 461–503.

—— (1957). *Mendelism and Evolution* (6th ed). Methuen, London.

—— (1964). *Ecological Genetics*. Methuen, London.

—— (1969). Ecological genetics. In *Scientific Thought 1900–1960* (ed. R. Harré), pp. 173–95. Clarendon Press, Oxford.

Fryer, J. C. F. (1913). An investigation by pedigree breeding into the polymorphism of *Papilio polytes*, Linn. *Philos. Trans. Roy. Soc. Lond. Ser. B* **204**, 227–54.

Goldschmidt, R. B. (1945). Mimetic polymorphism, a controversial chapter of Darwinism. *Q. Rev. Biol.* **20**, 147–64, 205–30.

Gould, S. J. (1983). The hardening of the modern synthesis. In *Dimensions of Darwinism. Themes and Counterthemes in Twentieth Century Evolutionary Theory* (ed. M. Grene), pp. 71–93. Cambridge University Press, New York.

Grene, M. (1974). Statistics and selection. In *The Understanding of Nature. Essays in the Philosophy of Biology* (ed. M. Grene), pp. 154–71. D. Reidel, Dordrecht. (Originally published 1961, *Br. J. Philos. Sci.* **9**, 110–27, 185–93.)

Haldane, J. B. S. (1932). *The Causes of Evolution*. Longmans, Green, London.

—— (1954). The statics of evolution. In *Evolution as a Process* (eds J. S. Huxley, A. C. Hardy & E. B. Ford), pp. 109–21. Allen and Unwin, London.

Hamilton, W. D. (1963). The evolution of altruistic behavior. *Am. Nat.* **97**, 31–3.

Hoyle, F. and Wickramasinghe, N. C. (1981). *Evolution from Space*. J. M. Dent, London.

Kettlewell, H. B. D. (1955). Selection experiments on industrial melanism in the Lepidoptera. *Heredity*, **9**, 323–42.

Kimler, W. C. (1983a). Mimicry: views of naturalists and ecologists before the modern synthesis. In *Dimensions of Darwinism. Themes and Counterthemes in Twentieth Century Evolutionary Theory* (ed. M. Grene), pp. 97–127. Cambridge University Press, New York.

—— (1983b). *One Hundred Years of Mimicry: History of an Evolutionary Examplar*. PhD thesis, Cornell University, Ithaca N.Y.

Kottler, M. (1980). Darwin, Wallace, and the origin of sexual dimorphism. *Proc. Am. Philos. Soc.* **124**, 203–26.

Lewontin, R. C. (1972). Review of *Ecological Genetics and Evolution, Essays in Honour of E. B. Ford* (ed. E. R. Creed). *Nature*, **236**, 181–2.

Marshall, G. A. K. (1908). On diaposematism, with reference to some limitations of the Müllerian hypothesis of mimicry. *Trans. Entomol. Soc. Lond.* **1908**, 93–142.

Maynard Smith, J. (1978). *The Evolution of Sex*. Cambridge University Press, London.

—— (1983). Current controversies in evolutionary biology. In *Dimensions of Darwinism. Themes and Counterthemes in Twentieth Century Evolutionary Theory* (ed. M. Grene), pp. 273–86. Cambridge University Press, New York.

Nicholson, A. J. (1927). A new theory of mimicry in insects. *Aust. Zoologist* **5**, 10–104.

Norton, B. (1983). Fisher's entrance into evolutionary science: the role of eugenics. In *Dimensions of Darwinism. Themes and Counterthemes in Twentieth Century Evolutionary Theory* (ed. M. Grene), pp. 19–29. Cambridge University Press, New York.

Olby, R. (1981). La théorie génetique de la selection naturelle vue par un historien. *Rev. Synthèse, IIIe S.*, **103–4**, 251–89.

Poulton, E. B. (1908). *Essays on Evolution 1889–1907*. Clarendon Press, Oxford.

—— (1912). Darwin and Bergson on the interpretation of evolution. *Bedrock*, **1**(1), 48–65.

Provine, W. B. (1971). *The Origins of Theoretical Population Genetics*. University of Chicago Press, Chicago and London.

—— (1985). The R. A. Fisher – Sewall Wright controversy and its influence upon modern evolutionary biology. *Oxford Surv. Evolut. Biol.* **2**, 197–200.

Punnett, R. C. (1905). *Mendelism* [1st edn]. Macmillan and Bowes, Cambridge.

—— (1911). *Mendelism* (3rd edn). Macmillan, London.

—— (1915). *Mimicry in Butterflies*. Cambridge University Press, Cambridge.

—— (1919). *Mendelism* (5th edn). Macmillan, London.

—— (1927). *Mendelism* (7th edn). Macmillan, London.

Sheppard, P. M. (1962). Some aspects of the geography, genetics, and taxonomy of a butterfly. In *Taxonomy and Geography* (ed. D. Nichols), pp. 135–52. Systematics Association, London.

—— Turner, J. R. G., Brown, K. S., Benson, W. W., and Singer, M. C. (1985). Genetics and the evolution of muellerian mimicry in *Heliconius* butterflies. *Philos. Trans. Roy. Soc. Lond. Ser. B*, **308**, 433–607.

Turner, J. R. G. (1977). Butterfly mimicry: the genetical evolution of an adaptation. *Evolut. Biol.* **10**, 163–26.

—— (1983). 'The hypothesis that explains mimetic resemblance explains evolution': the gradualist-saltationist schism. In *Dimensions of Darwinism. Themes and Counterthemes in Twentieth Century Evolutionary Theory* (ed. M. Grene), pp. 129–69. Cambridge University Press, New York.

—— (1985). Random genetic drift, R. A. Fisher, and the Oxford School of Ecological Genetics. In *The Probabilistic Revolution. Vol. 2: Ideas in the Sciences* (eds G. Gigerenzer, L. Krüger, and M. Morgan). MIT Press/Bradford Books, Cambridge, Mass. (in press).

—— (1986). *Mimicry in Butterflies*. Princeton University Press, Princeton. (In prep.)

Wallace, A. R. (1865). On the phenomena of variation and geographical distribution as illustrated by the *Papilionidae* of the Malayan region. *Trans. Linn. Soc. Lond.* **25**, 1–71.

—— (1889). *Darwinism. An Exposition of the Theory of Natural Selection with Some of its Applications* (2nd edn). Macmillan, London.

The R. A. Fisher–Sewall Wright controversy

WILLIAM B. PROVINE

Introduction

The intense controversy between R. A. Fisher and Sewall Wright, which lasted from 1929 until Fisher's death in 1962, was both highly visible and very influential in modern evolutionary biology. The controversy between Fisher and Wright has had a more fundamental and lasting impact upon evolutionary biology than any other controversy or rivalry in this century. Indeed, the controversy did not end with Fisher's death. Since 1962, Wright has produced a steady stream of papers and his four volume *Evolution and the Genetics of Populations* (1968, 1969, 1977, 1978a) in which he constantly contrasts his interpretations with those of Fisher. Many others have joined the battle on both sides.

The aim of this paper is to elucidate the crucial issues in the Fisher–Wright controversy and to sort them out from the huge number of anecdotal stories about the controversy. I also will show how the controversy exerted such a major influence upon evolutionary theory and field research. I have drawn freely from my forthcoming biography of Sewall Wright (Provine 1986), to which the reader is referred for greater detail. The complete correspondence between Fisher and Wright is reproduced in the biography; an excellent selection of the correspondence with critical commentary may be found in Bennett (1983). All of Wright's correspondence will soon be available for scholarly investigation at the Library of the American Philosophical Society in Philadelphia, as Fisher's is already available at the University of Adelaide.

First acquaintance

Fisher and Wright first met in the summer of 1924. Fisher had read and been impressed by Wright's series of papers, 'Systems of Mating' (1921a–e) in which Wright had applied his method of path coefficients to the problem of inbreeding and its implications for breeding theory and evolution (the method of path coefficients is a technique for quantifying a given path of causation, from cause to effect, in a complex causal scheme; it is particularly effective when applied to definite causal lines and linear relations as in Mendelian inheritance). Fisher attended the International Mathematical Congress in Toronto in the summer of 1924, after which he travelled to Washington DC where Wright was working at the Animal Husbandry Division of the United States Department of Agriculture. Wright was aware of Fisher's 1918 paper, 'The correlation between relatives on the supposition of Mendelian inheritance', and his reputation as a bio-statistician. At the time Fisher was working at the Rothamsted Experimental Station as a statistician for plant breeders and Wright was serving in a similar role for animal breeders at the USDA. So they had a great deal to share and were eager to meet each other.

Wright and Fisher had a long conversation about animal and plant breeding, statistical techniques, and the quantitative consequences of Mendelian heredity. Wright gave Fisher copies of several of his papers on path coefficients and animal breeding, and in return Fisher promised to send Wright a copy of his 1922 paper, 'On the dominance ratio', which Wright had never seen. The paper arrived on schedule and Wright found it very stimulating. This initial interchange between Wright and Fisher started a chain of events that led from occasional exchange of Christmas cards to, beginning in 1929, serious correspondence about quantitative evolutionary theory.

Fisher, Wright, and mathematical population genetics

Contrary to the opinion of many evolutionists, the Fisher–Wright controversy did not stem in any substantive way from disagreements in mathematical population genetics. On all occasions when Fisher and Wright appeared to disagree on quantitative questions, they were able to settle the differences and reach near total agreement.

Fisher and Wright used different quantitative approaches in their mathematical models of evolutionary change. Fisher favoured the differential and integral calculus, using continuous functions to build his models. Wright, however, used his method of path coefficients, which was particularly well adapted to the calculation of the effects of inbreeding in a finite population. Moreover, the quantitative approach of each was well suited to the theory of evolution each advocated. Thus, Fisher's approach was perfect for developing his fundamental theorem of natural selection and Wright's approach for his shifting balance theory of evolution in nature, in which inbreeding played a major role in the production of heritable variation between populations.

There can be no doubt that the same data sets can be interpreted very differently by using different quantitative approaches. The intense arguments between statisticians and biometricians over the best quantitative techniques to use in particular circumstances illustrates this point. Yet in the case of controversies between Fisher and Wright, the differences in their quantitative methods do not appear to explain their fundamental differences on questions concerning the evolutionary process. Their different quantitative methods led to almost exactly the same numerical results regarding the quantitative effects of inbreeding, mutation, migration, selection, and other variables, upon the statistical distribution of genes in a population.

Differences of the early years

Fisher was the first to have the idea of modelling the evolutionary process upon the changes in a statistical distribution of genes in a population. This was his great invention in the important mistitled 1922 paper, 'On the dominance ratio' (Fisher 1922). In his now famous 1918 paper, Fisher had examined the statistical consequences of such variables as genic interaction (epistasis), dominance, assortative mating, multiple alleles, and linkage upon the correlations between relatives. In the 1922 paper he extended the analysis to examine the effects of these and other variables upon the statistical distribution of genes in the population. Fisher was proposing that the changes in this distribution constituted the evolutionary history of the population.

The fundamental thoughts underlying many of the later quantitative works of Fisher and Wright on evolution were contained in this seminal 1922 paper. Beginning with the Hardy–Weinberg equilibrium, the distribution of genes expected under ideal assumptions, Fisher analysed the influence of selection, dominance, heterozygote advantage, mutation rate, random extinction of genes, and assortative mating upon the distribution of genes in the population. He showed, for example, that a single locus with two alleles, selection favouring one homozygote would lead to the elimination of the other allele. If, on the other hand, selection favoured the heterozygote, then the result was a stable equilibrium of the distribution of the alleles in the population. He also demonstrated that the survival at low frequency in a population of a rare mutation depended more upon chance than selection. Thus, large populations, in which a mutation would occur more often in absolute numbers, would be expected to maintain a higher genetic variability than small populations. This was a very exciting paper for those few who could understand what Fisher had done.

Wright did not have an extensive background in mathematics, not even in differential and integral calculus. He could not follow all of Fisher's differential equations and derivations in the 1922 paper, but he certainly did understand Fisher's attempt to incorporate the effects of crucial variables into one statistical distribution of genes in a population. Indeed, Wright was so impressed that within a year he was hard at work developing his own version of the statistical distribution of genes, using his method of path coefficients rather than the differential equations used by Fisher.

As Wright developed his model, he discovered that it differed in some significant and puzzling ways from that of Fisher. When Wright used his method of path coefficients to calculate the rate of decrease of heterozygosis in a population under no selection or mutation he obtained the figure $1/2N$, where N was the effective population size. Using differential equations, Fisher had arrived at the corresponding figure for the rate of decay of allele frequency as $1/4N$ (Fisher 1922, p. 330). Also, the factors in Wright's distribution for selection and dominance did not agree with Fisher's model, although Wright was not confident in these cases that his own method led to reliable results. Wright wrote up his paper on evolution and had it typed in the late fall of 1925, just as he was preparing to leave the USDA for a professorship at the University of Chicago. Despite his strong desire to submit the paper for publication, Wright held back because of the discrepancies with Fisher's model and his hope to push beyond it in some significant ways.

For the first 2 years at the University of Chicago, Wright was inundated with teaching duties and had little time to sort out the differences between his model and Fisher's. When he did attempt this on a few occasions, he could not work out the reasons for the differences in the models. Then, in 1928, Fisher published two papers on the evolution of dominance (Fisher 1928a,b). For reasons to be discussed later in this paper, Wright disagreed strongly with Fisher's theory of the evolution of dominance and wrote a rebuttal (Wright 1929a). Fisher thought that Wright disagreed with his calculations of the effects of small selection rates on gene frequencies over long periods and wrote a letter to Wright that said in part, 'What I mainly want to know . . . is whether you agree with me that a very slight selective effect acting for a correspondingly long time will be equivalent to a much greater effect acting for a proportionately shorter time' (Fisher to Wright, 6 June 1929). Wright replied that he was not challenging Fisher's mathematical calculation of the effects of selection in changing gene frequencies, but instead his conception of the evolutionary process (Wright to Fisher, 28 June 1929). Thus, in his published rebuttal to Wright's 1929 paper cited above, Fisher began by stating that Wright's 'primary formulas differ in no essential respects from my own and that the selective intensity which inclines Professor Wright to reject the theory is in fact the same that originally led me to adopt it' (Fisher 1929, p. 553). Indeed, on the one quantitative issue where they did disagree (the value of the selection pressure upon the heterozygote), Fisher was able to show that Wright's calculations were incorrect, but when corrected were in full agreement with his own calculations. In other words, they had no disagreement whatsoever on the quantitative models themselves with regard to the evolution of dominance.

By August 1929, Wright had rewritten his paper on evolution and he sent a copy to Fisher with the explicit hope that Fisher could discover the reason for the discrepancy in their figures for the rate of decay of heterozygosis in finite populations ($1/4N$ for Fisher, $1/2N$ for Wright). Fisher found the mistake in his earlier derivation, thanking Wright profusely in a letter for pointing out the discrepancy and adding that 'with this correction I find myself in entire agreement with your value $2N$ for the time of relaxation and with your corrected distribution for factors in the absence of selection' (Fisher to Wright, 15 October 1929). Fisher entered these changes in proof of his forthcoming book *The Genetical Theory of Natural Selection* (Fisher 1930).

However, their terms for selection in the stochastic distribution were still not identical. Wright noticed this particularly when Fisher sent him a copy of the book. Wright wrote to Fisher asking for a more detailed derivation of his selection term, a request to which Fisher responded immediately. With Fisher's derivation in hand, Wright was able to rederive his own selection term so that it gave results identical with those of Fisher. Wright entered this new derivation in both the proofs of his review of Fisher's book for the *Journal of Heredity* (Wright 1930) and in the proofs of his paper on evolution then in press at *Genetics* (Wright 1931). When Wright wrote to Fisher to tell him of these developments, Fisher responded, 'I am glad to hear that the little discrepancies are clearing themselves up' (Fisher to Wright, 25 October 1930). In his review of the *Genetical Theory of Natural Selection*, after an account of his correspondence with Fisher, Wright stated clearly that 'our mathematical results on the distribution of gene frequencies are now in complete agreement as far as comparable, although based on very different methods of attack' (Wright 1930, p. 352).

Only once after this full agreement of quantitative results in the early 1930s did Fisher and Wright appear to reach different conclusions as a consequence of different quantitative approaches. This concerned their disagreement about evolution in the rare species, *Oenothera organensis*, found only in the Organ Mountains of southern New Mexico [for an account of this controversy, see Provine (1986, Chapter 13)]. In response to Wright's analysis of the data (Wright 1939), Fisher responded in the second edition of *Genetical Theory of Natural Selection* (1958) with his own analysis using differential equations. He concluded that his quantitative results were 'very different' from those of Wright in 1939, and he attributed the differences to Wright's 'failure to develop any explicit formulae' and his reliance upon 'extensive numerical calculations based upon trial values of numerous constants he introduces' (Fisher 1958, p. 109). Wright answered with a careful comparison of the practical consequences of his and Fisher's models, concluding convincingly that they were virtually identical.

Thus, the differences in quantitative approaches of Fisher and Wright did not entail significant differences in their evolutionary theories. Nor was the tension between their quantitative approaches a creative force in evolutionary biology. Most modern evolutionists use whatever quantitative approach best suits the needs of the project at hand, and switch back and forth freely between Fisherian and Wrightian models. Frequently, these switches may be found in a single paper or chapter. The really important and influential tensions between Fisher and Wright had almost nothing to do with differences in the mathematical methods each employed.

Personal differences

Anywhere theoretical population genetics is discussed stories about the personality differences between Fisher and Wright abound. The rather few occasions when they were thrown together after the early 1930s, for example at Iowa State University, North Carolina State University, and the University of Cambridge, all generated gossip about how poorly they communicated in person. All correspondence between them, except for the unimpeded exchange of published papers, ceased in 1932. There were differences between them in both temperament and personality. In 1964, J. B. S. Haldane described these differences by saying that Fisher, when alive, 'preferred attack to defense. Wright is one of the gentlest men I have ever met and if he defends himself, will not counter-attack' (Haldane 1964, p. 344). I think that Haldane's assessment, although reflecting widely held opinions of evolutionists about Fisher and Wright, was misleading at best. Fisher probably did prefer attack to defence, but he also could have a strong defensive reaction, particularly when he thought the criticism or attack upon him was wrong or unwarranted [see the biography of Fisher by his daughter, Joan Fisher Box (1978)]. Haldane's characterization of Wright was just plain wrong. Wright was very shy, but was not gentle with his critics. How Haldane could have believed that Wright would not counter-attack is a mystery to me. He was actually the inveterate master of the complete counter-attack, frequently repeated many times over. Wright has always fully answered his critics.

A conflict was bound to result from these differences in personality. Wright continued criticizing Fisher, particularly his theory of the evolution of dominance, well after Fisher thought he had rebutted Wright's argument and after he had suggested to Wright in a letter that they cease to publish on their differences on the evolution of dominance. In 1933, however, Wright felt that he had never had the opportunity to really express in print his whole theory of the evolution of dominance. Thus, when Wright's paper on the evolution of dominance appeared (1934a), one section of which repeated his critique of Fisher's theory, Fisher wrote a very angry rebuttal (1934), which was followed by a strongly worded reaction from Wright (1934b). They were never on friendly terms again.

Although these personal differences were real and were certainly related to the tone of their disagreements over the years, and added a certain theatrical air to the conflict, I do not think that these differences go far enough in explaining the really important aspects of their conflict. Important scientists often have conflicts of personality, but such conflicts do not frequently have an important effect upon the advancement of the field. Even if Fisher and Wright had been the best of personal friends and had disagreed amicably, as many friends do, their disagreements on evolutionary theory would have been substantive and influential.

Evolutionary theory

FISHER'S THEORY

Qualitative evolutionary theory was the subject on which Fisher and Wright disagreed so strongly, and to such great effect. From his early student days at Cambridge University, Fisher was a devoted and self-conscious neo-Darwinian. By this, I mean that his general view of the evolutionary process fit closely with that of E. B. Poulton, E. Ray Lankester, Raphael Meldola, Karl Jordan, Leonard Darwin, and others who believed that natural selection was at all levels by far the most pervasive and important mechanism of evolution in nature. Under this view, every character of an organism had been shaped by the action of natural selection, whether or not biologists of the day had been able to spot the utility of the character. Combined with Mendelian heredity, understanding of the origin of variation, and recombination, Darwin's theory of natural selection was to Fisher's mind the nearly complete mechanism of evolution in nature.

Fisher was also greatly interested in mathematics, astronomy, and physics, and he found especially attractive the deep explanatory power of simple quantitative laws such as the inverse square law of gravitational attraction, Boyle's gas laws, and the second law of thermodynamics. He wanted to find the correspondingly simple quantitative law that would allow evolutionary phenomena to fall in place. His first attempt in this direction was in his 1922 paper, 'On the dominance ratio', in which he argued that an equation representing the stochastic distribution of Mendelian determinants in a population over time was the key to an accurate and quantitative understanding of evolution in that population.

Among many simplifying assumptions in the 1922 paper, Fisher assumed his population was extremely large and consequently had high storage of genetic variability. In such a population, he was able to demonstrate that his stochastic distribution led to the certain conclusion that of all the variables (selection, dominance, mutation rate, random extinction of genes, assortative mating, and epistasis), natural selection acting upon single genes was the supreme determinant of the evolutionary process. A mutation rate far higher than any observed in nature could be balanced by a tiny selection rate against it. Epistasis and random extinction were negligible. Fisher likened the stochastic distribution of genes, dominated by natural selection, to the general laws of the behaviour of gases.

> The investigation of natural selection may be compared to the analytic treatment of the Theory of Gases, in which it is possible to make the most varied assumptions as to the accidental circumstances, and even the essential nature of the individual molecules, and yet to develop the general laws as to the behaviour of gases, leaving but a few fundamental constants to be determined by experiment (1922, pp. 321–2).

However, the whole stochastic distribution was a rather messy equation, especially as more and more variables were incorporated into it. Thus, by 1930 Fisher had reached the view that the law was the real counterpart of the great laws of the physical sciences was his 'fundamental theorem of natural selection', which stated in words said, 'The rate of increase in fitness of any organism at any time is equal to its genetic variance in fitness at that time' (Fisher 1930, p. 35). After deriving and explaining the theorem, Fisher wrote:

> Professor Eddington has recently remarked that 'The law that entropy always increases – the second law of thermodynamics – holds, I think, the supreme position among the laws of nature'. It is not a little instructive that so similar a law should hold the supreme position among the biological sciences (Fisher 1930, pp. 36–7).

A fundamental aspect of Fisher's approach to biology was his expectation that the 'laws' of biology would be similar to the laws of physics.

WRIGHT'S THEORY

Wright's theory of evolution in nature was far more strongly modelled upon biology than was Fisher's. Wright built directly upon his extensive knowledge of experimental laboratory genetics, and the experience of animal and plant breeders. Wright also drew upon and was influenced by a very different natural history tradition than the neo-Darwinian one with which Fisher associated himself.

Four major research projects were influential in shaping Wright's theory of evolution in nature: (1) William Castle's selection experiment upon hooded rats; (2) Wright's thesis research on the interaction effects in guinea-pigs of the Mendelian factors determining colour and hair direction; (3) Wright's work on inbreeding, outbreeding, and selection in guinea-pigs; and (4) the analysis of the transformation of the Shorthorn breed of cattle during its foundation period (Wright 1978b).

From Castle's selection experiment upon hooded rats Wright learned two very important, but quite different points: that direct mass selection (selection in a random breeding population) was a powerful means of genetically changing the expression of a character, and that mass selection also had built-in limitations. The limitations had long been known to professional animal and plant breeders. Severe mass selection could indeed rapidly change the expression of a character such as milk production, but at the extreme and often unacceptable cost of deleterious side effects, distressingly expressed to breeders as loss of fecundity. In the breeding of large animals such as cattle or horses, mass selection was a slow and frustrating process, especially when the characters being selected had a low heritability (as was frequently the case). Fisher was also much impressed by Castle's selection experiments with hooded rats and cited this work frequently, especially in connection with his theory of the evolution of dominance. But Fisher emphasized only the positive effects of mass selection as had Darwin, a rather curious view considering that Darwin was highly knowledgeable of both animal and plant breeding, and Fisher was working at the Rothamsted Station at the time he published his theory of the evolution of dominance. If anything, Wright was more impressed by the limitations of mass selection.

Wright's thesis research upon the interaction effects of colour factors in guinea-pigs demonstrated that organisms were built up of complex interaction systems of genes rather than being, as Wright frequently emphasized, a mere mosaic of unit characters each determined by a single gene. From this fact of interaction, Wright deduced that to the breeder or under natural selection the selective process would be most effective if it operated upon interactive systems of genes rather than upon single genes. In a large random breeding population, the possible distinctive interaction systems of genes were rarely phenotypically expressed and thus not exposed to selection. As Fisher had shown clearly, in a large random breeding population selection was effectively limited to mass selection of single genes. Wright believed that both breeders and natural selection had more effective mechanisms than mere mass selection.

A clue to this more effective mechanism came from Wright's work on highly inbred strains of guinea-pigs at USDA. Because of the random fixation of genes caused by the many generations of inbreeding (brother-sister mating), each strain or family became fixed with a highly distinctive (almost) homozygous genetic complement. The inbreeding process had revealed the interaction systems so well hidden in the random breeding control population, making them available for the selection process. Of course, breeders would have to use intermediate rather than severe inbreeding to avoid the general decline in vigour and fecundity that usually followed intense inbreeding.

From his analysis of the origin of the Shorthorn breed of cattle, Wright discovered that breed had indeed undergone a time of intense inbreeding during its foundation period. Selection was applied to the variability revealed by the inbreeding. Diffusion from a very few herds to many others was accomplished by use of only a relatively few closely related sires, thus making over the entire breed. Wright thought that mass selection had played a relatively minor role.

Reasoning from his theory of optimal animal breeding to his theory of evolution in nature (as had Charles Darwin), Wright deduced that for natural selection to be a really efficient process, populations in nature must be subdivided into partially isolated subgroups small enough to cause the random drifting of genes and consequent manifestation of many different interaction systems. Optimally, the subgroups were large enough to prevent direct fixation by random drift, because this would lead to degeneration and extinction. Mass selection within subgroups followed by selective diffusion from particularly successful subgroups were the steps required for the transformation of the whole species.

IMPORTANCE OF TRADITIONS IN NATURAL HISTORY AND TAXONOMY

The other factor that must be kept in mind for understanding Wright's early expressions of his theory of evolution in nature is the natural history and taxonomic tradition that he followed, in contrast to the neo-Darwinian, wholly adaptationist tradition that Fisher followed. Right in Darwin's own work, there was a basic tension between adaptive and non-adaptive mechanisms of the evolutionary process (for explanation and documentation see Provine 1985a). Both traditions were very active and generating controversy in the last two decades of the nineteenth century and first two decades of the twentieth century. By the time Wright was formulating his theory of evolution in nature, there was a very strong tradition in both taxonomy and natural history that challenged the neo-Darwinian view. Represented by such figures as Moritz Wagner, John T. Gulick, David Starr Jordan, G. J. Romanes, Vernon L. Kellogg, H. E. Crampton, F. B. Sumner, A. C. Kinsey, G. C. Robson, and O. W. Richards, followers of this tradition argued that many, if not most, of the differences observed between closely-related species were of no adaptive value whatsoever and such differences must have arisen from some other mechanism than natural selection. Geographical isolation was always raised as a contributing factor to this non-adaptive differentiation.

Wright clearly believed that his theory of population subdivision could explain both the adaptive and the non-adaptive differences, depending upon the degree of subdivision and consequent effect of random genetic drift, whereas the neo-Darwinian view could explain non-adaptive differences only by the rather weak explanation that such characters were linked with others of such high adaptive value that their combination was adaptive (the theory of correlation).

Population structure and traditions in taxonomy and natural history are therefore the keys to understanding the basic differences between the early evolutionary views of Fisher and Wright. Fisher's view, based upon very little evidence, was that evolution proceeded by the mass selection of single genes in very large random breeding populations. This view of the evolutionary process was impossible if natural populations were subdivided the way that Wright thought. Wright's belief, also based upon very little evidence from natural populations, was that the assumption of large random breeding populations was unwarranted and that appropriately subdivided populations could lead to a better understanding of both artificial and natural selection. But Wright's view of the evolutionary process was impossible if Fisher was correct in his assumption about natural populations being large and random breeding. Although Wright could have no quarrel with Fisher's fundamental theorem of natural selection given Fisher's assumptions, he did think that the theorem, to be accurate, should be stated as follows: 'The rate of increase in fitness of any population at any time is equal to its genetic variance in fitness at that time, *except as affected by mutation, migration, change of environment, and effects of random sampling*' (Wright to Fisher, 3 February 1931). The tension between these views was deep and inevitable, and moreover could not be settled by any amount of theorizing, no matter how quantitative. Only careful study of natural populations could resolve the tension.

The controversies

In the space available here I will be able to examine briefly only three of the many controversies in which Fisher and Wright engaged, but these should give sufficient insight to see their significance. These three controversies were over: (1) the evolution of dominance, (2) the general mechanism of evolution in nature, and (3) evolution in the moth *Panaxia dominula*.

THE EVOLUTION OF DOMINANCE

Fisher applied his basic evolutionary theory first to the evolution of mimicry (Fisher 1927) and then to the evolution of dominance (Fisher

1928a,b). His thesis was that these apparent cases of evolution by discontinuous leap were in fact the result of deterministic small selection pressures acting over long periods of time upon very small heritable modifiers always available in large random breeding populations. Noting that most observed mutations in *Drosophila melanogaster* were wholly recessive, Fisher offered the following explanation. Most mutations, he argued, were recurrent, deleterious, and occurred at a finite rate. The mutations that geneticists had observed in the laboratory in *Drosophila* must also have occurred in nature. Since natural populations were large and random breeding, a deleterious mutant allele was likely to become fixed by reaching a state of equilibrium between adverse selection and recurrent mutation. Natural selection would tend to make the heterozygote and mutant homozygote phenotypically resemble the homozygous wild type. Castle's experiment on hooded rats had shown how selection could accumulate modifiers of a mutant to change its appearance very significantly. Thus dominance was not an immutable property of the gene. Because heterozygotes were vastly more numerous in the population than the mutant homozygotes, natural selection would over time make the heterozygotes resemble the homozygous wild type, thus accomplishing dominance. A recessive allele was a potential wild type allele and the force required for the change was a minute selective advantage. Fisher acknowledged that the selective advantages causing the evolution of dominance were extremely small: 'It may be calculated that with mutation rates of the order of one in a million, the corresponding selection in the state of nature, though extremely slow, cannot be safely neglected in the case of the heterozygotes' (Fisher 1928a, p. 126).

Wright objected strongly to Fisher's theory of the evolution of dominance on two grounds (Wright 1929a). First, he objected to Fisher's whole theory of the mechanism of evolution in nature. He thought Fisher's theory ignored factors that, in his opinion, would swamp the tiny selection rates operating over long periods of time that were hypothesized in Fisher's theory of the evolution of dominance. Primary among these factors were random genetic drift from sampling effects in relatively small populations and the selective pressures caused by interactive effects of the rest of the genome. Second, he had reason to believe from his research in physiological genetics that recessivity generally resulted from inactivation of the gene, thus reducing the product produced by the dominant allele. Interestingly, in his first published reply to Fisher's theory of the evolution of dominance, Wright did not even mention small effective population sizes or random genetic drift. That was because he had not yet published his long paper on evolution in which he treated, in some detail, the issue of population size and random drift, and he did not see a reasonable way to introduce this complex issue in a brief note. Thus, in the first published interchange between Wright and Fisher, it was impossible to detect that what was really at issue was two basically different theories of evolution in nature.

In his published reply to Wright's criticism of his theory, Fisher even used selectively neutral genes to buttress his mass selectionist viewpoint:

As to ratios having neutral stability, there is one reason for thinking that the factors suffering the feeblest selective action will at any one time be the most numerous. The fate of those powerfully selected is quickly settled; they do not long contribute to the variance. It is the idlers that make the crowd, and very slight attractions may determine their drift. On the whole, it seems that the most reasonable assumption which we can make, on an obscure subject, is that the effect is approximately equal to the cause (Fisher 1929, p. 556).

Wright was genuinely surprised to find Fisher using the existence of nearly neutral genes to support his intensely selectionist view of the evolutionary process (what a contrast with Ohta and Kimura!). This time Wright felt compelled to use his arguments of effective population size and random drift in his published answer to Fisher, but when he sent the draft of this answer to Fisher, he also sent the manuscript of his long paper on evolution. Thus, in addition to wanting to clear up the inconsistencies in their quantitative models, Wright wanted to provide Fisher with the background to one of his strongest objections to Fisher's theory of the evolution of dominance.

In his published statement, Wright said that he could not understand how Fisher could use the existence of almost neutral factors to argue for the all prevailing power of natural selection. Indeed, Wright asked, how small did the effective breeding size (N) of a population have to be before random drift would swamp the effects of selection rates of the size postulated by Fisher?

Unfortunately it is difficult to estimate N in animal and plant populations. In the calculations, it refers to a population breeding at random, a condition not realized in natural species as wholes. In most cases, random interbreeding is more or less restricted to small localities. These and other conditions such as violent seasonal oscillation in numbers may well reduce N to moderate size, which for the present purpose may be taken as anything less than a million. If mutation rate is of the order of one in a million per locus, an interbreeding group of less than a million can show little effect of selection of the type which Dr. Fisher postulates even though there be no more important selection process and time be unlimited (Wright 1929b, p. 560).

Wright clearly believed that natural species were not random breeding populations. Instead, he believed that random breeding was 'more or less restricted to small localities' over the entire range of the species. Thus, the effective breeding size of a whole species was vastly less than the number of breeding individuals in the species.

Fisher's reply to this argument in a letter to Wright is very instructive.

I am not sure that I agree with you as to the magnitude of the population number N. To reduce it to the number in a district requires that there shall be *no* diffusions even over the number of generations considered. For the relevant purpose I believe N must usually be the total population on the planet, enumerated at sexual maturity, and at the minimum of the annual or other periodic fluctuation. For birds, twice the number of nests would be good. I am glad, however, that you stress the importance of this number (Fisher to Wright, 13 August 1929).

By this time, both Fisher and Wright knew that the real issue between them was not merely a disagreement about the evolution of dominance, but a deep disagreement about the evolutionary process in general.

Their disagreement over the evolution of dominance immediately became a subject of great interest to evolutionists. J. B. S. Haldane, H. J. Muller, C. R. Plunkett, E. M. East, and others immediately joined the controversy and interest in it has never died away. Almost every decade the question of the evolution of dominance is 'definitively' settled by one or another evolutionist, only to emerge again as a difficult issue. A good review up to 1978 may be found in Wright's *Evolution and the Genetics of Populations*, Volume 3 (1977), but much has appeared on the evolution of dominance since then.

GENERAL THEORY OF EVOLUTION IN NATURE

The controversy over the evolution of dominance obviously raised for Fisher and Wright awareness of the differences in their general theories of the evolutionary process. However, widespread appreciation of the tension between their evolutionary theories came only with the publication of their major works between 1930 and 1932. Fisher's *Genetical Theory of Natural Selection* (appeared May 1930) was the first major work to explore in sophisticated quantitative detail the synthesis of Mendelian genetics with evolutionary theory and is a landmark in the history of twentieth century evolutionary biology. Here Fisher developed in detail his view that evolution in nature occurred in large random breeding populations in which the overwhelming factor determining changes in gene frequencies was natural selection acting upon individual genes. Wright's contrasting view of the evolutionary process can best be seen in his review of Fisher's book (Wright 1930), his big paper 'Evolution in Mendelian Populations' (Wright 1931), and in his paper delivered at the Sixth International Congress of Genetics, 'The roles of mutation, inbreeding, crossbreeding, and selection in evolution' (Wright 1932).

In the review Wright presented the contrasting view that he found most appealing. Instead of the large panmictic population emphasized by Fisher, he argued that,

> A much more favourable condition would be that of a large population, broken up into imperfectly isolated local strains. . . . The rate of evolutionary change depends primarily on the balance between the effective size of population in the local strain (N) and the amount of interchange of individuals with the species as a whole (m) and is therefore not limited by mutation rates. The consequence would seem to be a rapid differentiation of local strains, in itself non-adaptive, but permitting selective increase or decrease of the numbers in different strains and thus leading to relatively rapid adaptive advance of the species as a whole. Thus, I would hold that a condition of subdivision of the species is important in evolution not merely as an occasional precursor of fission, but also as an essential factor in its evolution as a single group (Wright 1930, pp. 354–5).

This was the first clear statement of what Wright later termed his 'three phase shifting balance' theory of evolution, involving large subdivided populations, random drift, intrademe, and interdeme selection. This was a clear alternative to Fisher's whole theory of evolution in nature.

The general impression of the differences between the theories of Wright and Fisher did not, however, reflect the full sophistication of either theory. I do not think that the impact of Fisher's theory of the inevitable deterioration of the environment, which caused a continual change in fitnesses and therefore in the parameters of his fundamental theorem, was well appreciated in the early years. Nor was Wright's shifting balance theory widely understood or appreciated. Instead, for understandable reasons, the conflict between their evolutionary theories became seen in the 1930s and, for the most part, later as one between Fisher's pan-selectionism and Wright's random genetic drift. Wright himself initially neglected to take into account the deterioration of the environment in evaluating Fisher's theory, although he did in the 1932 paper after Fisher complained in a letter (Fisher to Wright, 19 January 1931). For his part, Fisher never relinquished the view that Wright was advocating the importance of straight random genetic drift as an important mechanism of evolution, whereas Wright always argued that in his shifting balance theory random drift merely provided the variation upon which natural selection then acted to provide adaptive advance.

To be sure, there was much room for confusion about Wright's shifting balance theory of evolution. Since the late 1940s, Wright has consistently denied that he ever attributed any important role to random drift, except as a mechanism for generating variability upon which selection then acted. Thus, in 1967 Wright stated:

> Many critics have seized on the concept of random drift that was proposed and have asserted that I have advocated this as a significant *alternative* to natural selection. Actually, I have never attributed any evolutionary significance to random drift except as a trigger that may release selection toward a higher selective peak through accidental crossing of a threshold (Wright 1967, p. 254–5).

And more recently in 1982, Wright declared: 'I emphasize here that while I have attributed great importance to random drift in small local populations as providing material for natural selection among interaction systems, I have never attributed importance to non-adaptive differentiation of species' (Wright 1982, p. 12). These statements must be compared with what Wright actually said in the years 1929–1932.

(1) The non-adaptive nature of the differences which usually seem to characterize local races, subspecies, and even species of the same genus indicates that this factor of isolation is in fact of first importance in the evolutionary origin of such groups, a point on which field naturalists (e.g., Wagner, Gulick, Jordan, Osborn, and Crampton) have long insisted (Wright 1929b, pp. 560–1).

(2) The actual differences among natural geographical races and subspecies are to a large extent of the non-adaptive sort expected from random drifting apart (Wright 1931, p. 127).

(3) Fisher's theory is one of complete and direct control by natural selection while I attribute greatest immediate importance to the effects of incomplete isolation (Wright 1931, p. 149 f.n.).

(4) The direction of evolution of the species as a whole will be closely responsive to the prevailing conditions, orthogenetic as long as these are constant, but changing with sufficiently long continued environmental change (Wright 1931, p. 151).

(5) Adaptive orthogenetic advances for moderate periods of geologic time, a winding course in the long run, non-adaptive branching following isolation as the usual mode of origin of subspecies, species, perhaps even genera, adaptive branching giving rise occasionally to species which may originate new families, orders, etc., . . . are all in harmony with this interpretation (Wright 1931, p. 153).

(6) Under the shifting balance process complete isolation originates new species differing for the most part in non-adaptive respects, but is capable of initiating an adaptive radiation as well as of parallel orthogenetic lines, in accordance with the conditions (Wright 1931, p. 158).

(7) Complete isolation of a portion of a species should result relatively rapidly in specific differentiation, and one that is not necessarily adaptive. The effective intergroup competition leading to adaptive advance may be between species rather than races. Such isolation is doubtless usually geographic in character at the outset, but may be clinched by the development of hybrid sterility (Wright 1932, p. 363).

(8) That evolution involves non-adaptive differentiation to a large extent at the subspecies and even the species level is indicated by the kinds of differences by which such groups are actually distinguished by systematists. It is only at the subfamily and family levels that clearcut adaptive differences become the rule (Robson 1928; Jacot 1932). The principal evolutionary mechanism in the origin of species must then be an essentially non-adaptive one (Wright 1932, pp. 363–4).

(9) Subdivision into numerous local races whose differences are largely non-adaptive has been recorded in other organisms wherever a sufficiently detailed study has been made. [There follows citation of the work of Gulick, Crampton, David Starr Jordan, Ruthven, Kellogg, Osgood, Kinsey, Osborn, Rensch, Schmidt, David Thompson, and Sumner] (Wright 1932, pp. 364–5).

Viewed all together at one time, these citations illuminate the question of why Wright's shifting balance theory was so misunderstood in the 1930s and later. The careful reader of Wright's papers in 1932 would almost certainly conclude that non-adaptive random drift following isolation was a primary mechanism in the origin of races, subspecies, species, and perhaps genera. One can easily understand why Fisher and other biologists understood Wright to be saying that random drift was an important mechanism alternative to selection in the origin of subspecies and species. Yet, at the same time, one can understand why Wright insists with reason that he has always argued that evolution in nature depends upon a balance of forces, of which random drift is only one. In the early 1930s, however, Wright correctly understood the taxonomists and naturalists to be telling him that most of the differences between closely related species were non-adaptive. Thus, he set the 'balance' in his shifting balance theory to give room for differentiation at the species level from random drift. Later, after the 1940s, when systematists led by Ernst Mayr and David Lack argued that non-adaptive differences between species were rare and adaptive ones the rule, Wright naturally saw his shifting balance theory as applicable.

CONTROVERSY OVER *PANAXIA DOMINULA*

Ever since the beginning of evolutionary biology, conspicuous polymorphisms in natural populations had presented evolutionists with a serious problem. How could a single primary mechanism of evolution, whether natural selection, inheritance of acquired characters of even an orthogenetic force lead to conspicuous dimorphisms within a single interbreeding population? All naturalists were familiar with at least some cases of such dimorphism. An obvious, but important reason why the debates over explaining the origin of conspicuous polymorphisms have been so persistent is that, until the rise of molecular biology, conspicuous polymorphisms were the most easily accessible (often it seemed, the only) measurable heritable characteristics. Thus, conspicuous polymorphisms have been constantly in the forefront of research on evolution in natural populations.

Darwin concluded in the *Origin* that natural selection could not be the explanation of polymorphic species (Darwin 1859). Indeed, when he defined the concept of natural selection in chapter IV of the *Origin*, Darwin specifically dissociated natural selection from polymorphism:

> This preservation of favourable variations and the rejection of injurious variations, I call Natural Selection. Variations neither useful nor injurious would not be affected by natural selection, and would be left a fluctuating element, as perhaps we see in the species called polymorphic (Darwin 1859, p. 81).

Conspicuous polymorphisms were the focus of disagreement about adaptive versus non-adaptive evolution from Darwin's day until the 1940s, when the prevailing view of evolutionists became strongly adaptationist. But before that, the prevailing view was almost as strongly non-adaptationist, with the obvious exception of the extreme neo-Darwinians such as Poulton, and of course later Fisher and Ford. Evolutionists today have mostly forgotten that as recently as the early 1940s, most evolutionists believed, along with Darwin, that conspicuous polymorphisms were non-adaptive. Any evolutionist today knows that Mayr considers the vast majority of conspicuous polymorphisms to be adaptations shaped by natural selection. Yet, consider what Mayr says in his *Systematics and the Origin of Species* (1942),

> Neutral polymorphism is due to the action of alleles "approximately neutral as regards survival value". Ford (following Fisher) believes that this kind of polymorphism is relatively rare, because "the balance of advantage between a gene and its allelomorph must be extraordinarily exact in order to be effectively neutral". This reasoning may be correct in all the cases in which one of the alternative features has a definite survival value or at least is genetically linked with one. There is, however, considerable indirect evidence that most of the characters that are involved in polymorphism are completely neutral, as far as

survival value is concerned. There is, for example, no reason to believe that the presence or absence of a band on a snail shell would be a noticeable selective advantage or disadvantage. Among the many species of birds which occur in several clear-cut colour phases (Stresemann 1926 and later papers), there is, with one or two exceptions, no evidence for selective mating or any other advantage of any of the phases.

Even more convincing proof for the selective neutrality of the alternating characters is evidenced by the constancy of the proportions of the different variants in one populations. The most striking case is that of the snails *Cepaea nemoralis* and *C. hortensis*, in which Diver (1929) found that the proportions of the various forms from Pleistocene deposits agree closely with those in colonies living today (Mayr 1942, p. 75).

Fisher and Ford strongly disliked this view of conspicuous polymorphisms in the 1930s and 1940s, and thought that Sewall Wright was the theoretician who had provided the modern justification, namely random drift, for such an interpretation. To combat Wright and this view, Fisher and Ford began their research on *Panaxia dominula* and Ford began the ambitious research programme that later provided the evidential basis for his monumental *Ecological Genetics* (1964 and later editions).

The details of the debate over *Panaxia dominula*, fascinating though they are, are too complicated for inclusion here (for a full account see Provine 1986, Chapter 12). The crux of the debate, however, is clear enough. Fisher and Ford carefully followed one population of the day-flying moth *Panaxia dominula* for the years 1941–1946, using the marking, release, and recapture method to generate data from which gene frequencies and effective population sizes could be estimated. They studied an easily observed polymorphism controlled by simple Mendelian inheritance, with the added convenience that all three Mendelian classes could be distinguished by sight. They found that the yearly fluctuations in gene frequency were too large to be caused by random genetic drift in a population of the size they had measured. They concluded, by elimination, that the observed changes in gene frequency must have resulted from fluctuations in natural selection. They concluded,

> Thus our analysis, the first in which the relative parts played by random survival and selection in a wild population can be tested, does not support the view that chance fluctuations in gene ratios, such as may occur in very small isolated populations, can be of any significance in evolution (Fisher and Ford 1947, p. 173).

The denial by Fisher and Ford of *any* significance for random genetic drift in evolution on the basis of just one experiment spurred Wright to answer their challenge. Wright's defence of random drift took two very different forms, one of them new and characteristic of his future attitude toward the evolution of conspicuous polymorphisms. First, he challenged the adequacy of their data to support the conclusion that random drift could not be the cause of the observed fluctuations in gene frequency. Second, he used the very different argument that, even if the observed fluctuations were caused by fluctuations in natural selection, this held only for a single locus with two alleles in a case of conspicuous polymorphism and it did not follow that all genes varied in frequency for that same reason. 'The situation is similar', Wright argued, 'except for the element of intent, to one that is familiar to livestock breeders. With very intensive selection for particular characters, others must be allowed to vary at random if numbers are to be maintained' (Wright 1948, p. 285).

The debate over *Panaxia dominula* intensified, with angry exchanges on both sides (Fisher and Ford 1950; Wright 1951). More than anything else, the debate crystallized the differences between the Fisherian and Wrightian ways of thinking about evolution in nature. Each side was now even more motivated to produce supporting field research.

The effect upon Wright was to make him rethink the whole issue of conspicuous polymorphisms. Even before he became aware of the results of Cain and Sheppard, strongly indicating that conspicuous polymorphisms in colour and banding of the shell in *Cepaea nemoralis* were subject to strong selection pressures, Wright had already sent a letter to Cain arguing that conspicuous polymorphisms in general should be expected to have evolved under and remain under strong selective forces. Instead, he argued, conspicuous polymorphisms were a very small proportion of the genome where random drift would play almost no role; however, with the rest of the genome, random drift might be an important factor (Wright to Cain, 14 November 1950). Also, by the late 1940s, Wright had dropped his earlier view that random drift could cause non-adaptive differences between species. Random drift was still crucially important, generating novel interaction systems at the level of the local semi-isolated population. However, the action of natural selection, he argued, would cause such differences as might be observed even at the subspecies level to be adaptive. Thus, Wright fits well into the shift towards a more selectionist attitude that Stephen Jay Gould aptly describes as the 'hardening of the synthesis' in the late 1940s and early 1950s (Gould 1983).

Influence of the Fisher–Wright controversy upon evolutionary biology

In this section I will support my earlier assertion that the controversy between Fisher and Wright had a great influence upon modern evolutionary biology. I should say immediately, however, that by no means all of the influence of Wright and Fisher came from their disagreements. Together with Haldane, Hogben, Chetverikov, and other quantitative evolutionists, Fisher and Wright had a very important joint influence upon evolutionary biology (Provine 1978). All the mathematical populationists agreed upon a number of specific points, such as the immense power of selection to change gene frequencies in an intuitively surprising small number of generations, the relative insignificance of mutation pressure in relation to selection pressure under most conditions, or the theory of balanced polymorphisms that flowed from heterozygote advantage. They also agreed, for the most part, upon which variables were the really important ones, such as selection rates, effective population size, or population structure. Finally, the work of the population geneticists was a crucial element in the vast narrowing of the controversies over the mechanisms of evolution in nature.

The evolutionary synthesis of the 1930s and 1940s certainly did not remove all controversy about mechanisms of microevolution or speciation, but what it did do, and resoundingly, was to greatly narrow the range of controversies that existed before 1930. An evolutionist like Henry Fairfield Osborn was a very prestigious man before the synthesis, but who now talks about 'Aristogenesis' or about any of the host of other theories that were so common and taken seriously by one or another major school of thought before 1930? The mathematical population geneticists played a central role in this narrowing of the possible mechanisms of evolution, primarily by demonstrating quantitatively that some mechanisms were not as powerful as they seemed intuitively, and others were totally superfluous. Within this narrowed scope of the mechanisms of evolution in nature, the controversy between Fisher and Wright did have an important and specifiable impact.

GENERAL INFLUENCE OF THE CONTROVERSY

The general influence of the Fisher–Wright controversy during the period of the evolutionary synthesis is deeply related to the long-standing debate about adaptive and non-adaptive mechanisms of evolution in nature. This issue has consistently attracted much attention from evolutionary biologists, from Darwin's day to the present (for a detailed account of the ongoing controversy about adaptive versus non-adaptive mechanisms of evolution, see Provine 1985a). During the period of the synthesis, the debate often became focused as 'Wright's concept of random drift versus Fisher's concept of natural selection', although Wright himself was advocating his shifting balance theory rather than simply random drift as an alternative to selection. Most of the major books and papers on general evolutionary theory during the synthesis period reflected the tension between either random drift versus selection or shifting balance theory versus selection in large random breeding populations. The latter was more faithful to the ideas of Wright and Fisher, but the former was more prevalent.

The tension between Fisher and Wright may easily be seen in such works as (in chronological order) Ford's *Mendelism and Evolution* (1931), Dobzhansky's *Genetics and the Origin of Species* (1937), Huxley's *The New Systematics* (1940) and *Evolution: the Modern Synthesis* (1942), Mayr's *Systematics and the Origin of Species* (1942), Simpson's *Tempo and Mode in Evolution* (1944) and Stebbins' *Variation and Evolution in Plants* (1950), all of which were influential works during the synthesis period. Although rather few young evolutionary biologists during the 1930s and 1940s read the technical papers of Fisher and Wright, they could not help being familiar with the tension between their views of evolution from the more accessible literature. Thus, the tension between Fisher and Wright was interwoven into the very fabric of evolutionary theory during this period.

FIELD RESEARCH ON GENETICS AND EVOLUTION
IN NATURAL POPULATIONS

Fisher and Wright agreed on the centrality of certain variables in the evolutionary process, among them effective population size, selection rates, and population structure, but they disagreed strongly on the relative sizes of many of the variables they agreed were important. It was precisely the agreement on the crucial variables combined with disagreement on the sizes of the variables that provided such a great stimulus to field research on the genetics of natural populations. Theoretical population genetics could not settle the questions about sizes of variables; that required field research.

Before the early 1930s, those who studied natural populations (such as F. B. Sumner with *Peromyscus*), had few handles to guide their studies. But the pertinent variables of microevolution were clear by the mid-1930s. If a field researcher could only determine, for example, effective population size, then this would constitute a base for distinguishing between a Wrightian and a Fisherian pattern of evolution in the organism. Or if observing the frequency of a gene from year to year or season to season could be used in combination with measures of effective population size to estimate the relative roles of selection and random drift. Dobzhansky, Ford, Huxley, Timoféeff-Ressovsky, and others sounded the clarion call to use the variables pinpointed by the mathematical population geneticists to guide studies of natural populations that could in turn be used to discriminate between the models of evolutionary change proposed by the same population geneticists.

> The experimental work that should test these mathematical deductions is still in the future, and the data that are necessary for the determination of even the most important constants in this field are wholly lacking. Nonetheless, the results of the mathematical work are highly important, since they have helped to state clearly the problems that must be attacked experimentally if progress is to be made. . . . The manner of action of selection has been dealt with only theoretically, by means of mathematical analysis. The results of this theoretical work (Haldane, Fisher, Wright) are, however, invaluable as a guide for any future experimental attack on the problem (Dobzhansky 1937, p. 121, p. 176).

> 'The work of Fisher, Haldane, and Wright is of the greatest importance, showing us the relative efficacy of various evolutionary factors under the different conditions possible within the populations. It does not, however, tell us anything about the real conditions in nature, or the actual empirical values of the coefficients of mutation, selection, or isolation. It is the task of the immediate future to discover the order of magnitude of these coefficients in free-living populations of different plants and animals; this should form the aim and content of an empirical population genetics' (Timoféeff-Ressovsky 1940, p. 104).

Both Dobzhansky and Timoféeff-Ressovsky were hoping primarily to discriminate between the evolutionary schemes of Fisher and Wright.

The field researches that began as attempts to discriminate between the evolutionary theories of Fisher and Wright are among the most important of the evolutionary synthesis period. No detailed analysis is possible here (for that see Provine, 1985b, 1986), but the primary studies on the genetics of natural populations of which I am speaking are Dobzhansky's 'Genetics of natural populations' series (Wright collaborated on five of the first fifteen; Lewontin *et al*. 1981), the collaboration of Fisher and Ford on *Panaxia dominula* (Fisher and Ford 1947), of Cain and Sheppard on *Cepaea nemoralis* (1950), the work of Lamotte on *Cepaea nemoralis* in France (Lamotte 1951), and the ambitious, but abortive attempts of Buzzati-Traverso *et al*. to study the genetics of natural populations in Italy (1938). All of these except the last were ongoing projects, some continuing to the present day and still reflecting their origin in the controversy between Fisher and Wright.

One way to see graphically the great impact the tension between Fisher and Wright had upon the study of genetics and evolution in natural populations is to examine the first edition of Ford's *Ecological Genetics* (or the later editions). Even a cursory reading will reveal that most of the field research described in the book was begun with the controversy between Fisher and Wright explicitly in mind. The book is therefore in one sense a monument to the productive stimulus that the controversy provided.

Conclusions

Controversy in science frequently produces nothing but harsh words, hurt feelings, unpleasantries, and backbiting. Yet at other times, controversies manifesting many of these same symptoms have very positive and stimulating effects upon the whole field, and are central to later developments. The controversy between Fisher and Wright was one of these.

Indeed, the controversy between Fisher and Wright was in my opinion so central to modern evolutionary biology that it has become invisible to young people in the field. I always recommend that any aspiring graduate student in the field of evolutionary biology read Wright's four volume *Evolution and the Genetics of Populations* (in reverse order!), the second edition of Fisher's *Genetical Theory of Natural Selection*, and his papers on evolution from the *Collected Papers of R. A. Fisher* (Bennett 1971–1974). Those who complete this exercise are invariably much impressed by how alive and central are many of the issues upon which Wright and Fisher disagreed to current evolutionary theory and research on natural populations.

References

Bennett, J. H. (ed.) (1971–1974). *Collected papers of R. A. Fisher*. The University of Adelaide, Adelaide.
—— (ed.) (1983). *Natural Selection, Heredity, and Eugenics*. Clarendon Press, Oxford.

Buzzati-Traverso, A., Jucci, C., and Timoféeff-Ressovsky, N. W. (1938). Genetica di popolazioni. *Consiglio Nazionale Delle Ricerche, La Ricerva Scientifica, Series II,* **1**, 3–30.

Cain, A. J. and Sheppard, P. M. (1950). Selection in the polymorphic land snail *Cepaea nemoralis* (L.). *Heredity,* **4**, 275–94.

Darwin, C. R. (1859). *On the Origin of Species.* John Murray, London.

Diver, C. (1929). Fossil records of Mendelian units. *Nature,* **124**, 183.

Dobzhansky, T. (1937). *Genetics and the Origin of Species.* Columbia University Press, New York.

Fisher, R. A. (1918). The correlation between relatives on the supposition of Mendelian inheritance. *Trans. Roy. Soc. Edin.* **52**, 399–433.

—— (1922). On the dominance ratio. *Proc. Roy. Soc. of Edinburgh,* **42**, 321–41.

—— (1927). On some objections to mimicry theory – statistical and genetic. *Trans. Roy. Entomol. Soc. Lond.* **75**, 269–78.

—— (1928a). The possible modification of the response of the wild type to recurrent mutation. *Am. Nat.* **62**, 115–26.

—— (1928b). Two further notes on the origin of dominance. *Am. Nat.* **62**, 571–4.

—— (1929). The evolution of dominance: a reply to Professor Sewall Wright. *Am. Nat.* **63**, 553–6.

—— (1930). *The Genetical Theory of Natural Selection.* Clarendon Press, Oxford. Second edition, 1958.

—— (1934). Professor Wright on the theory of dominance. *Am. Nat.* **68**, 370–4.

—— and Ford, E. B. (1947). The spread of a gene in natural conditions in a colony of the moth *Panaxia dominula. Heredity,* **1**, 143–74.

—— and —— (1950). The 'Sewall Wright effect'. *Heredity,* **4**, 117–9.

Fisher Box, J. (1978). *R. A. Fisher: The Life of a Scientist.* Wiley, New York.

Ford, E. B. (1931). *Mendelism and Evolution.* Methuen, London.

—— (1964). *Ecological Genetics.* Methuen, London.

Gould, S. J. (1983). The hardening of the modern synthesis. In *Dimensions of Darwinism* (ed. M. Grene), pp. 71–93. Cambridge University Press, Cambridge.

Haldane, J. B. S. (1964). A defense of beanbag genetics. *Persp. Mod. Biol. Med.* **7**, 343–60.

Huxley, J. S. (ed.) (1940). *The New Systematics.* Oxford University Press, Oxford.

—— (1942). *Evolution: the Modern Synthesis.* Allen and Unwin, London.

Jacot, A. P. (1932). The status of the species and the genus. *American Naturalist,* **66**, 346–64.

Lamotte, M. (1951). *Rescherches sur la structure génétique des populations naturelles de Cepaea nemoralis L.* Supplement au *Bulletin Biologique de France et de Belgique,* No. 35.

Lewontin, R. C., Moore, J. A., Provine, W. B., and Wallace, B. (1981). *Dobzhansky's Genetics of Natural Populations.* Columbia University Press, New York.

Mayr, E. (1942). *Systematics and the Origin of Species.* Columbia University Press, New York.

Provine, W. B. (1978). The role of mathematical population geneticists in the evolutionary synthesis of the 1930s and 1940s. *Stud. Hist. Biol.* **2**, 167–92.

—— (1985a). Adaptation and mechanisms of evolution after Darwin: a study in persistent controversies. In *The Darwinian Heritage* (ed. D. Kohn), pp. 825–66. Princeton, Princeton University Press, Wellington.

—— (1985b). The study of the genetics of natural populations during the evolutionary synthesis of the 1930s and 1940s. In *La vita e la sua storia* (ed. L. Bullini, M. Ferraguti, F. Mondella, and A. Oliverio), pp. 121–8. *Scientia.*

—— (1986). *Sewall Wright: Geneticist and Evolutionist*. University of Chicago Press, Chicago (in press).

Robson, G. C. (1928). *The species problem*. Oliver and Boyd, London.

Simpson, G. G. (1944). *Tempo and Mode in Evolution*. Columbia University Press, New York.

Stebbins, G. L. (1950). *Variation and Evolution in Plants*. Columbia University Press, New York.

Stresemann, E. (1926). Uebersicht über die "Mutationsstudien" I–XXIV und ihre wichtigsten Ergibuisse. *J. Ornith.* **74**, 377–385.

Timotéeff-Ressovsky, N. W. (1940). Mutations and geographical variation. In *The New Systematics* (ed. J. S. Huxley), pp. 73–136. Clarendon Press, Oxford.

Wright, S. (1921a). Systems of mating. I. The biometric relation between parent and offspring. *Genetics*, **6**, 111–23.

—— (1921b). Systems of mating. II. The effects of inbreeding on the genetic composition of a population. *Genetics*, **6**, 124–43.

—— (1921c). Systems of mating. III. Assortative mating based on somatic resemblance. *Genetics*, **6**, 144–61.

—— (1921d). Systems of mating. IV. The effects of selection. *Genetics*, **6**, 162–6.

—— (1921e). Systems of mating. V. General considerations. *Genetics*, **6**, 168–78.

—— (1929a). Fisher's theory of dominance. *Am. Nat.* **63**, 274–9.

—— (1929b). The evolution of dominance. *Am. Nat.* **63**, 556–61.

—— (1930). *The Genetical Theory of Natural Selection*, by R. A. Fisher (review). *J. Hered.* **21**, 349–56.

—— (1931). Evolution in Mendelian populations. *Genetics*, **6**, 97–159.

—— (1932). The roles of mutation, inbreeding, crossbreeding and selection in evolution. *Proc. 6th Int. Cong. Genetics*, **1**, 356–66.

—— (1934a). Physiological and evolutionary theories of dominance. *Am. Nat.* **68**, 25–53.

—— (1934b). Professor Fisher on the theory of dominance. *Am. Nat.* **68**, 562–5.

—— (1939). The distribution of self-sterility alleles in populations. *Genetics*, **24**, 538–52.

—— (1948). On the roles of directed and random changes in gene frequency in the genetics of populations. *Evolution* **2**, 279–94.

—— (1951). Fisher and Ford on the Sewall Wright effect. *Am. Scient.* **39**, 452–8, 479.

—— (1967). The foundations of population genetics. In *Heritage from Mendel* (ed. R. Alexander Brink), pp. 245–63. University of Wisconsin Press, Madison.

—— (1968). *Evolution and the Genetics of Populations. Vol. 1. Genetic and Biometric Foundations*. The University of Chicago Press, Chicago.

—— (1969). *Evolution and the Genetics of Populations. Vol. 2. The Theory of Gene Frequencies*. The University of Chicago Press, Chicago.

—— (1977). *Evolution and the Genetics of Populations. Vol. 3. Experimental Results and Evolutionary Deductions*. University of Chicago Press, Chicago.

—— (1978a). *Evolution and the Genetics of Populations. Vol. 4. Variability within and among Natural Populations*. University of Chicago Press, Chicago.

—— (1978b). The relation of livestock breeding to theories of evolution. *J. Anim. Sci.* **46**, 1192–200.

—— (1982). The shifting balance theory and macroevolution. *Ann. Rev. Genetics*, **16**, 1–19.

Essay review: the relationship between development and evolution

KEITH STEWART THOMSON

Evolution and Development. J. T. Bonner (ed.). Springer-Verlag, Berlin, Heidelberg and New York. 1980. 357 pp.
Development and Evolution. B. C. Goodwin, N. Holder, and C. C. Wylie (eds). Cambridge University Press, Cambridge. 1983. 437 pp.
Embryos, Genes and Evolution. R. A. Raff and T. C. Kaufman. Macmillan, New York. 1983. 395 pp.

The question of the relationship between development and evolution is a bit like the statement (mistakenly) attributed to Mark Twain: 'everybody talks about the weather, but nobody does anything about it'. There is a great deal of talk about a new entry of developmental biology into the study of evolution. How much is actually being done about it?

There has, of course, been a long historical tradition of interest in the relationship of ontogeny to phylogeny. This was essentially a matter of attempting to discover process indirectly from analysis of pattern. de Beer (1940) and Gould (1977) have kept alive the phenomenon of heterochrony as something more, namely a mechanism for change. Present efforts to relate development and evolution are, however, concerned with fundamental processes in development, all the way from molecular genetics to morphogenesis. The publication of three important new books offers the chance to review this exciting new field.

Current interest in the relationship of development to evolution stems from two sources: a dissatisfaction with the modern versions of Darwinism evolution growing out of the New Synthesis of the 1940's, and a resurgence of interest in the cellular and morphogenetic (rather than just molecular and genetic) aspects of developmental biology. While modern evolutionary studies include many triumphs, some obstinately recurrent old questions are still not answered. These questions form a focus for a renewed interest in so-called macroevolution. Trends, correlation of morphological change (both at the point of origin of major groups and in their subsequent evolution), variable rates of evolution, and different rhythms or modes of evolution, are not much closer to solution now than they were in 1944 when Simpson's *Tempo and Mode of Evolution*, or in 1947 when Jepsen *et al.'s Genetics, Paleontology and Evolution* ushered in the great expansion of post-war evolutionary studies.

All scientific problems, essentially by definition, are about cause. Darwin's great contribution to the study of evolution was to tackle the problem in terms of causes that could be isolated and studied mechanistically. As Hodge (1977) has shown, his whole method was based on Hershel's treatment of *vera causa* (1830). Indeed, Hodge argued that each chapter of the Origin is arranged according to the three-fold content of *vera causa*: (1)

the existence of a causal process, (2) the competence of that process to produce the effect to be explained, and (3) demonstration of the responsibility of the projected mechanism for that effect, which then make it a possible source of explanation of other phenomena. In the last analysis Darwin did not fully establish cause, the basic insoluble problem being the proof of historical hypotheses. One can establish elements one and two, but not the third.

Like all sciences, evolutionary biology has become the art of the possible or, as Medawar rephrased it, the Art of the Soluble. Not surprisingly, therefore, it has become preoccupied with what it is possible to define and measure, namely with selection, with mathematical theories of population genetics, and with field and laboratory studies of populations. Also, the theories and methods of molecular genetics have added the possibility of getting at some fundamental mechanisms acting within the genome. The power of what is possible has produced the danger of our falling into the trap of assuming that all of evolution will eventually be deduced from (reduced to) theories of population genetics and molecular genetics.

The 'developmentalist approach' seeks to discover whether the processes of development add any significant component of causality in evolutionary mechanisms. This requires not only a detailed knowledge of developmental biology, but also a model of the components of an evolutionary theory in which such causality can be postulated and tested. Recent studies have provided the first such scheme.

The tradition of western science tends to make us relate theories and mechanisms in schemes of hierarchy. Formal hierarchical theory holds promise of becoming an important component in evolutionary biology (Vrba and Eldredge 1984; Salthe 1985). Hierarchical analysis allows us to analyse a very simple array of separate but interacting causes constituting an hypothesis of a comprehensive evolutionary mechanism. Such analysis depends first and foremost on the postulation of a series of focal levels at each of which comparable processes operate. In any evolutionary process, these fundamental elements are the introduction of variation at a given focal level and the sorting (i.e., both selective and non-selective processes) of that variation. The scheme that Vrba and Eldredge propose obviously derives a great deal from the 'units of selection' debate (Lewontin 1970; Hull 1980; see also Williams 1985 for review). Each focal level in the nested hierarchy is defined as having the properties of an individual (rather than a class; see discussion in Hull 1980) and has its own 'emergent' properties which are more than the sum of the parts of the lower levels. That is, each level has its own rules of operation and its own rules of 'assembly'. For example, the focal level 'deme' must be defined as more than simply 'a large number of individual organisms', it reflects instead a special set of interactions among those organisms, involving breeding systems, gene flow, and so on.

For our purposes, a very simple hierarchical outline of evolutionary mechanisms can be proposed in which the levels at which significant causation occurs are, at a minimum: genomic constituents, individual organisms, demes, species (see Vrba and Eldredge 1984). Vrba and

Eldredge would go further by proposing a level higher than 'species', but this introduces controversial elements not necessary to the present discussion (see Salthe 1985). Generally, the domain of traditional Neo-Darwinian evolution has been processes operating at the level of sorting of variation among individual organisms and the study of the differential properties of populations. More recently, species selection (Stanley 1975) has been proposed as a parallel set of mechanisms acting at the species level where the variation that is introduced is the variation among species, where presumably mutation rate is represented as speciation rate and where differential extinction rates represent sorting. At the other end of the hierarchy, we have the familiar concept of genic selection, that is sorting of variant constituents of the genome according to their 'fitness' in the general genetic context. There are also non-selective processes at the genomic level, such as the proliferation of 'selfish DNA'.

This then is an outline hypothesis into which we can fit other focal levels, if and when it is possible to show that there are other places (kin groups, for example) where a distinct element of cause is added to the system. In order to add such levels it is necessary to show that they are levels with their own rules, emergent properties and causes, and we can then attempt to discover these rules and their contribution to the whole.

The importance of such an hierarchical scheme to the present discussion is that it forces us to look for all the potential components of causality in evolutionary mechanisms. We then immediately find a major gap in our current knowledge. Reductionist Neo-Darwinian evolution has largely (but not wholly, of course) been pursued as though the mechanisms by which variation is introduced at the level of the organism – the creation of individual phenotypes – were insignificant. If only by default, variation introduced at the level of the individual organismal phenotype is commonly treated as if it were a simple mapping or variation at the genetic level. If this were really the case, genetics would give us a perfect read-out of individual variation and, presumably therefore, vice versa. But, as S. Kauffman, in his essay in Goodwin et al. (1983) (reviewed here), states, 'Only ignorance in general would have persuaded us that the response of any integrated complex dynamical system to random alterations in its parameters or structure would be fully isotropic'. The whole thrust of the developmentalist approach to evolution is to explore the possibility that asymmetries in the introduction of variation at the focal level of individual phenotypes, arising from the inherent properties of developing systems, constitutes a powerful source of causation in evolutionary change.

Goodwin (1982), in arguing for a more structuralist approach to evolution (see below), states that the central question in the relationship of development to evolution is becoming, 'what type of generative process may be operating in evolution to produce . . . transformational asymmetry'. The focus then shifts 'from a pre-occupation with the contingent and the historical . . . to concern with general principles of organization and transformation. . . . If a rational generative biology is ever realized, it will be based upon developmental processes as the logical, as well as the historical, origin of species' (p. 55).

In principle, developmental processes, which constitute the 'black box' between the properties of the genomic constituents and the properties of individual phenotypes, have the possibility, through contribution to the causation of variation at the level of the individual phenotype, to produce asymmetries in both (1) the nature of variation (i.e., what varies), and (2) the mode and rate of variation (i.e., how things vary). Such processes could then, in principle, contribute to the cause of evolutionary trends, of any kind of irregularity or 'saltatory' effect, phyletic constraints, correlations among changes within the single organism, stasis or very rapid change, multiple speciation phenomena (species flocks), and other phenomena loosely included in 'macroevolution' as well as effects on a microevolutionary scale. Indeed, the hierarchical scheme proposed above makes the terms macro- and microevolution superfluous.

Two simple examples will perhaps serve to show where an 'internalist' developmental approach to cause in evolution may help to explain problematic phenomena. The first involves the arrangement of the bones around the ankle joint in reptiles. In all primitive reptiles the astragalus and calcaneum were part of the foot, and the ankle joint passed proximal of them, between them and the distal ends of the tibia and fibula. In most advanced reptiles the astragalus and calcaneum are attached to the ends of the tibia and fibula, and the ankle joint is distal to them all. In crocodilians there is an oblique joint in which the astragalus is joined to the tibia and the calcaneum is part of the foot. It is difficult to see how a process of selection among adult phenotypes could effect the transition from one discrete, fully functional, pattern to another. There is no way in which these bones could move gradually across the joint to the other, with appropriate reorganization of all the muscle attachments, by means of a progressive transformation of a succession of adult phenotypes under strong selection. Nor is it reasonable to imagine the calcaneum first becoming reduced and lost from one side of the joint, and then slowly reappearing on the other side. Such a major structural innovation requires some sort of threshold effect (see Rachootin and Thomson 1981; Thomson 1982) in the reorganization of the rudiments of the foot at the limb bud stage of development. How could such a reorganization be caused? A single structural gene mutation is unlikely. However, a threshold effect, involving accumulation of changes in the basic parameters of a strongly canalized limb pattern, could equally be caused by strong selection, or by the build-up of mutations or recombinations without reference to any selective regime directly related to ankle function, but acting within the framework of internal structural rules.

A second example would be the evolution of the horse limb. The traditional explanation of the origin of oligosyndactyly in the horse limb would invoke strong selection for reduction of the lateral digits and elongation of the whole limb as an adaptation for fast running over hard terrain. This requires the generation of significant numbers of individuals within populations in which the lateral digits are very slightly reduced in size, leading to loss, presumably under random mutation and recombination, and perhaps aided by the effects of processes acting in small isolated populations. If there is a developmental input to the causation of such a set of changes it must be in developmental phenomena that cause a bias in the introduction of variation, a bias that may subsequently turn out to be adaptive. Such a bias must have its origins in the rules of limb development.

As noted in the introduction it is one thing to talk about possible mechanisms, and quite another to demonstrate both that they exist and that they have actually operated as causal factors in evolutionary change. However, without question, this subject is one of the important new frontiers in evolutionary biology.

The three books under review have very different approaches to the subject. Bonner (1982) is the proceedings of a Dahlem conference in which a number of developmentalists and evolutionists were brought together. In general the result is rather disappointing from the evolutionary point of view. The developmentalist authors produce interesting reviews of developmental data, but too often conclude that they were not quite sure what the question was. The evolutionists are more sure of what the question is, but although they are convinced that the answers must come from study of developmental constraints and the timing of developmental processes, they don't know what to do next. From this work one would have to conclude that the greatest hope for the future comes from the study of morphogenetic processes at the cell, tissue, and organ level, and that the reduction of developmental processes to molecular terms is a long way off, if it is possible at all.

Goodwin *et al.* (1983), from a symposium of the British Society for Developmental Biology, consists of a series of elegant essays by developmental biologists each putting their specialized field into the broadest comparative context and searching for common ground, together with a smaller number of theoretical contributions. It is an excellent survey of a whole range of fields at the morphogenetic level. By and large questions of evolutionary theory are not addressed, except in an introductory essay by Maynard Smith. A discussion of ontogeny and phylogeny by Patterson is couched in terms of systematic theory. There is a very heavy concentration on mechanisms of pattern formation in morphogenesis rather than molecular or cell-level processes.

Raff and Kaufman (1983) is different again in that it is the first attempt to produce a broad synthesis in text-book format of developmental data from the molecular to morphogenetic levels, aiming directly at some evolutionary questions. It is a stylish summary of the state of the art, that sticks closely to the fundamentals of what is actually known about development, plus some of the modern models. It is the more valuable for being written from a molecular and genetic viewpoint and I am sure it will turn out to be a landmark volume. No major evolutionary conclusions or speculations about mechanisms are drawn from the data and the only limitation of the work is that it concentrates too much upon explaining the phenomenon of heterochrony. Indeed, the first general conclusion that one reaches after reading these three books together is that, while heterochrony is an interesting phenomenon, it is peripheral to the central question and has captured too much attention among evolutionists. The real question involves the way in which fundamental processes of development respond to genetic and other perturbation, and vice versa. That heterochronic mechanisms involving size, shape, and time of sexual maturity can produce powerful effects is not in question, but it seems unlikely that this is a predominant mode of evolutionary change.

Taken together, these three books demonstrate both the high level of current interest in the possibilities of finding new developmental explanations for evolutionary phenomena and the difficulties that still stand in our way. The difficulties are both technical and conceptual. Discovering the causal relationship between development and evolution involves more than cataloguing a series of interesting developmental phenomena. After all, everything in the phenotype has a developmental history. If developmental processes have a potential role in the causation of evolutionary change (in addition to genome-level, population-level, or species-level phenomena) we need to understand those regularities, rules, and laws of developmental processes that constrain and direct the contingent phenomena of genetic change or environmental insult. These inherent (emergent) properties of development potentially add a realm of causality to the complex mechanism of evolutionary change, a causality that is not simply explained by (reducible to) the operation of genetic factors alone.

A simple example can be seen in the properties of epithelia (see also Thomson 1984). Some of the characteristics of epithelia can be reduced to the properties of their constituents. They consist of cells for example and the cells in epithelia show a special case of the 'rules' of cells in having a basement membrane. They may then be understood simply in terms of those lower level phenomena. However, other, emergent, properties of epithelia are independent of their cellular nature. They are not the properties of cells, but of sheets, and are properties held in common by sheets of paper, films of liquids, and layers of gases.

We can imagine an artificial model in which the folding of an epithelial sheet is due to a combination of change of cell shape and of buckling of the sheet of cells growing under mechanically constrained conditions. The control of folding of the sheet would then be in part genetic (in the control of cell division, cell shape, overall spatial geometry, properties of extracellular matrices, etc., in that individual) and also would result from contingent 'whole-organism' factors involving the growth of that particular epithelium in a given spatial context. However, control over the epithelium would also rest in the general rules of cell morphology (in the range of shapes that epithelial cells can assume), and in the rules of epithelial morphogenesis (in the ways that any sheet of cells can fold). These two sets of control points provide a definition of the 'causes' of a given pattern of epithelial folding. Therefore, in order to know how the shape of an epithelial organ can be changed (can evolve) we would need to understand a great deal more developmental biology than the suites of mechanisms by which gene expression is regulated and the ways in individual cells work under genetic instruction. It is obvious, for example, that in this model the rules of cellular morphology will exercise a control over change in cell shape that will override forces for change arising from a whole range of genetic 'instructions'. Similarly, the pattern of folding will, in part, be independent of variations in cell shape, and the contingencies of the spatial context may control whether folding occurs anyway. Further, a given pattern of epithelial folding could be produced by a broad range of different combinations of values in the four controlling levels, as long as each were held within some 'acceptable' range. The potential for evolutionary change in epithelial organs in this model would be a function of the potential for change in the system of genetic instructions, the inherent morphogenetic rules of cells and epithelia, whole-organism interactive contingencies, and the regime of sorting of realized phenotypes, as well as any factors acting at yet higher or lower hierarchical levels.

It is these sorts of rules and regularities in development, together with its epigenetic complexity, that produce the extremely important concept of 'developmental constraint', a key feature of all discussions of the relationship of development and evolution. The existence and possible evolutionary significance of such constraints have been studied particularly intensively in the vertebrate limb, as reviewed by Holder in Goodwin *et al.* (1983) and a recent article (Holder 1983). An interesting consequence of 'constraints' is the possibility that certain morphologies are 'forbidden', a proposition that, like so many in evolutionary biology, is impossible to 'prove'. However, developmental constraints are not necessarily only 'restrained'. If constraints in developmental mechanisms introduce asymmetries in the introduction of variation at the phenotypic level that are subsequently found to be adaptive, then such constraints can have a very powerful role in the causation of evolutionary change. S. Kauffman discusses the question briefly in his theoretical article in Goodwin *et al.* (1983) demonstrating a theoretical basis for orthogenesis. Such constraints would be a prime element in an internalist explanation of the evolution of the horse limb.

The evolutionist therefore needs to link up with developmental biology in order to master, not just the facts of particular pathways, but their rules and controls. Maynard Smith states, in his essay in Goodwin *et al.* (1983), that we need a theory of development comparable to the 'clear and highly articulated theory of evolution' that we already have. I would turn this around somewhat and say that we will not articulate a comprehensive theory of evolutionary mechanisms until we can include it in critical reference to an equally complete understanding of (a complex theory of) the causes and control of development.

Whence can we derive such an understanding of development? If the laws, rules, and regularities of development were strictly reducible to the laws of molecular genetics, then they could be deduced directly from the latter and would be uninteresting in the emergent sense. The examples just given show that such a simple reduction is not possible. From the evolutionary side there seem to be several indirect lines of enquiry. First, it should be possible to deduce some of the rules by the study of their consequences in the laws and regularities of phenotypic morphology. Here the work of Riedl (1978) and Seilacher (1970) essentially stand alone as serious modern attempts to develop an analytical methodology that will generate generalizations in the study of morphology. Secondly, the rules might be inferred, in part, from the study of variability in natural populations. For example, a recent study by Hanken (1984) of carpal and tarsal anomalies in the terrestrial salamander *Plethodon cinereus* aims to show that asymmetries in the morphological variants produced in a single isolated population are the result of developmental constraints in the process of limb pattern formation. Such conclusions are, of course, only justifiable if all alternate explanations (the direct expression of allele frequencies or patterns of recombination, for example) are ruled out, and if their cause can be directly linked to known factors in limb morphogenesis. One can come closer through a comparative study of developmental histories in related taxa, and by linking this to information on the underlying morphogenetic processes, and especially to the evidence that mutants provide concerning the capacity for, and rules and patterns of change. Such an approach has been followed extremely fruitfully in the work of Hinchliffe on the tetrapod limb (see Hinchliffe and Johnson 1982; Hinchliffe and Griffiths, in Goodwin *et al.* 1983). However, inevitably one returns to the need to understand the fundamentals of the nature and control of morphogenetic mechanisms in development.

Any developmental process can be thought of as including at least the following important sets of phenomena. (1) The rules of a set of basic parameters of the mechanism concerned. This might be the rate of cell division in a tissue level phenomenon, or some quantitative aspect of molecular assembly at the genetic level. (2) Control of spatial patterning, for example the redistribution of cytoplasmic determinants in a fertilized ascidian egg and the old mosaic-regulative dichotomy, or the apparent roles of the ZPA and AER in tetrapod limb buds. This category would include the range of phenomena underlying the 'oppositional information', 'polar co-ordinate', and 'induction' models for the explanation of pattern formation, and mathematical-structural regularities included in the concept of 'pre-pattern'. (3) Rules of structure and materials, which might be the rules of epithelial and mesenchymal behaviour in organ-level morphogenesis or of molecular structure. (4) Rules of structural and functional integrity. These rules must set the context for operation of the sets of contingent phenomena that affect species-, environment-, and time-specific parameters. The same sorts of rules should apply throughout all phases of development, from gene regulation and expression, gene function, the establishment of information, and control (pattern formation) to the mechanisms of cell interaction, morphogenesis, or the mechanisms of cell differentiation. How close are we either to discovering these rules and thereby to gaining direct information about the potential of developmental systems in evolutionary mechanisms or to modelling them usefully?

It is clear that there are still huge gaps in our understanding of the relationship between genetics and development. Models of gene regulation, expression, and function in eukaryotes are still incomplete. Therefore, in Bonner (1982), the group report on 'genomic change and morphological evolution' written by Dawid begins with a disclaimer: 'the consensus view emerged that present knowledge about genome function is not sufficient to make a large direct contribution. We do not know the mechanisms by which gene activity affects the development of an individual animal, therefore, we cannot come to useful specific conclusions regarding genomic correlates of evolutionary change at the morphological level'. Instead the group catalogued properties of genome organization as a basic starting point. Britten followed with a review of eukaryotic genome organization. Davidson, however, then presented two models of regulatory mechanisms with evolutionary potentials: repeated sequence and post-transcriptional control networks, and local multigene regulatory units.

It is frustrating to the evolutionary biologist that we still lack comparative data on even the most general temporal patterns of gene expression in animals or plants. Apart from the work summarized by Galau *et al.* (1976) comparing various stages of development in the sea urchin, we have little information even on the proportion of the genome that is active at different stages. The fact that a huge proportion of the genome in the sea urchin is active at the earliest stages and a relatively small proportion is switched on in adult differentiated tissues is tantalizing. It establishes that the 'housekeeping genes' in sea urchins are only a small proportion of the total genome. What does the pattern mean?

In amphibians, a considerable body of information has been built up concerning the pattern of activation of ribsomal RNAs in development, but less is known about the mRNAs and very few comparative data are available. This line of enquiry is of crucial importance because eventually it must lead to a comprehensive account of the mechanisms of very early pattern formation in embryos, and particularly the transfer from control of pattern in the egg by maternally-derived factors to control by the zygotic genome. A great deal of work remains to be done in this latter area, both experimentally and in the postulation of explanatory models. A basic review of the morphogenetic side of the subject is given in an excellent new book by Slack (1983). The more molecular side seems to be dominated by a strictly deterministic approach, a search for molecular cytoplasmic determinants in the egg, for example, rather than in the epigenetic approach that had dominated in previous decades (see commentary by Cooke and Webber 1983).

Little of the exciting work on the very earliest stages (unfertilized egg to the appearance of overt embryonic organization) is touched upon in Bonner (1982) and Goodwin *et al.* (1983), and relatively little in Raff and Kaufman (1983). Part of the reason must be that the field is still so unsettled, but it also reflects the basic premises (amounting to prejudice) that the early stages of development are most resistant to change, that any change that did occur would be bound to be lethal, and that, therefore, the interesting evolutionary effects must be produced at later morphogenetic stages. This is a view that owes its origin at least to von Baer's first and second laws. However, while the inadequacies of our information are incontrovertible, the fact remains that observable differences between major groups of organisms must have involved fundamental changes in early as well as late pattern formation and it seems extremely unlikely that all of these developmental differences have arisen from secondary reshuffling of development after gradual accumulation of a large mass of 'terminal' phenotypic change. It is an interesting question whether the differences between anuran and urodelan amphibians, with respect to the mechanics of gastrulation or the origin of the primordial germ cells, is secondary or directly related in some way to the origin of the groups. The 'invention' of the prechordal plate by vertebrates surely represents a primary restructuring of chordate gastrulation itself. Nieukoop and Sutarurya ('Some problems in the development and evolution of the chordates'), in Goodwin *et al.* (1983), engage in speculations about the role of basic restructurings in relationship to particular schemes of vertebrate phylogeny. However, it has to be admitted that it is a weak argument in favour of a phylogenetic hypothesis that some particular congruent developmental transition was 'possible'. Sander's paper in the same volume ('The evolution of patterning mechanisms: gleanings from insect embryogenesis and spermatogenesis') is somewhat more successful. Walbot ('Morphological and genomic variation in plants') and Frankel ('What are the developmental underpinnings of evolutionary changes in protozoan morphology?'), again in Goodwin *et al.* (1983), explore the possibility of major morphological changes having a relatively simple genetic basis.

At the Dahlem conference, edited by Bonner, cellular mechanisms in morphogenesis come in for considerable attention, but the 'group report' by Gerhardt once again was unable to formulate what the answer was 'to the question of the cellular basis of morphogenetic linkage'. The report by Gerhardt and also particularly the review by Wessells ('Processes responsible for metazoan morphogenesis') do, however, provide a valuable overview of some basic cell-level processes.

In these three books, especially those of Bonner and Goodwin *et al.* once one gets away from the molecular and very early developmental stages, a wholeheartedly epigeneticist view takes over. Hall and Horder (in Goodwin *et al.* 1983), and Maderson and Alberch (in Bonner 1982) are all explicit in viewing the epigenetic nature of development as a prime feature in its evolutionary potential. In this, these works follow directly in the tradition of Waddington (1975; cf. Rachootin and Thomson 1981) whose influence still dominates the whole subject. The essays by Hall ('Epigenetic control in development and evolution') and Horder ('Embryological bases of evolution') are particularly useful because they discuss some of the basic concepts and mechanisms involved in pattern formation and its control, a subject which has made solid advances with respect to the vertebrate limb, vertebrate hard tissues, and insect segmentation. It is here that the principal concepts of pattern control – tissue interactions (induction), positional information, polar co-ordinates, and prepattern (but see below) – are refined and tested.

In fact, there is no shortage of good reviews of mechanisms of morphogenesis and its control. To the three books reviewed here one must certainly add those of Slack (1983), Graham and Wareing (1984), and Hinchliffe and Johnson (1980). In Goodwin *et al.* (1983), French summarizes the 'Development and evolution of the insect segment', Maden, Gribbin, and Summerbell ('Axial organization in developing and regenerating limbs'), and Holder ('The vertebrate limb: patterns and constraints in development and evolution'), review the limb story (see also the essay by Hinchliffe and Griffiths). These essays summarize our knowledge of the basic morphogenetic and pattern forming mechanisms, although reference to the current issue of any major journal will show how the picture is changing all the time, especially in the field of homeotic mutants in *Drosphila*.

The aim of these essays is not just to describe the mechanisms, but also to demonstrate the properties of these systems that may have an evolutionary potential. Wolpert ('Constancy and change in the development and evolution of pattern'), in Goodwin *et al.* (1983), develops this theme. There is, however, a reciprocal incompleteness between the two major available systems. The control of insect segmentation is becoming understood at the genetic level, but the morphogenetic pattern control mechanisms are far less clear. Knowledge of the control of pattern in the vertebrate limb is far advanced, but its genetic basis is almost completely unknown. Other important systems, such as the role of the neural crest in vertebrate morphogenesis, somitogenesis in vertebrates, and most morphogenetic processes in invertebrates and plants, where it is possible to postulate a potential for the cause of evolutionary change, are similarly incompletely known. Thus, despite the optimism of the group report by Maderson ('The role of development in macroevolutionary change') in the Bonner volume, this potential remains tantalizingly difficult to translate into more concrete terms. As already noted, it is one thing to conclude that a phenomenon is possible, even likely, and another to show that it actually happened in the past or is operating now.

What more do we need? Of course, we need evolutionary biologists to use an increased knowledge of developmental biology and an expanded view of evolutionary mechanisms to define the critical questions about evolution itself, but we are still limited by the state of knowledge about development. It also seems to be the case that we need to break free from some of the traditional styles of evolutionary and developmental research. Goodwin has forcibly argued in a number of articles (Webster and Goodwin 1982; Goodwin 1982, 1984) that a 'structuralist' approach has great value in the search for the laws of biological form. S. Kauffman in Goodwin *et al.* (1983: 'Developmental constraints: internal factors in evolution') uses such an approach in reviewing the mathematical constraints that may explain plant phylotaxis, shell patterns in gastropods, and echinoid skeletal morphology. He also discusses combinatorial coding as an explanation of compartmentalization in insect segments and the properties of 'evolutionarily stable gene regulatory networks'. Goodwin and Trainor (in Goodwin *et al.* 1983) use a computer model to stimulate the 'generative principles' of the tetrapod limb. In this model, a field of values is set up and can be controlled by variation of a series of simple parameters. The model can thereby be caused to generate a range of different patterns mimicking given taxa, for example the four digit limb of

Necturus. As the authors point out, the traditional explanation of the four digit limb has involved the assumption that a preaxial or post-axial digit has been lost. However, if their generative model has any relationship to developmental reality, then a transformation from one limb pattern to another would not involve conservation of particular forms (digit 2, for example).

Such a generative scheme would have powerful potential for an internalist cause of evolutionary change (the horse limb, for example). It allows maintainance of the functional integrity of the limb. It can produce transformations without the needs for infinitely graded phenotypic transitions. It conforms with the predictions of developmental constraints in the limb with respect to the so-called forbidden morphologies. (It is, however, also a model in which such proscriptions form part of the premises, therefore one gets out what one puts in.) A similar analysis has been performed by Goodwin and Trainor for patterns of cleavage in embryogenesis (1983). The limitation of such modelling of the limb is that the parameters that generate the developmental field have not yet been expressed in terms of (or fully tested against) established facts of the genetic and cellular level control of limb development.

Other work on the structural basis of patterning mechanisms, such as Murray's famous paper on pelage patterns in mammals (1981) are centrally related to this line of enquiry; see also, for example, Odell *et al.* (1980). All of this brings us a long way from a more traditional concentration on allometry, heterochrony, and quantitative genetics, the limitations of which are discussed by S. Kauffman (*op. cit.*), but reinforces the point that the central questions of development that relate to evolution are the questions of control, and of the causes of lawfulness in development.

There are three important points arising from these sorts of discussion that require special mention. The first is that different ways of approaching problems should not be exclusive. The great value of the sort of hierarchical analysis mentioned here is that it demonstrates the relationships of mechanisms operating at different levels to the whole scheme of cause. Indeed, hierarchy theory predicts a whole range of upward and downward cascades of effect within a hierarchy that it was not necessary to discuss here. Further, an internalist view of causality in evolution should not deny the importance of the externalist view. The aim is not to replace one by the other, but rather than the two should be complementary. Similarly, a structuralist or field approach to generation of pattern in development should be complementary to and amplify the epigenetic approach. It is still a weakness that our knowledge of the basic facts of developmental mechanisms do not provide a way of linking the two in practice. A prime goal ought, for example, to be to express the sort of models that Goodwin and Trainor (and others) have created in terms of what we know about mechanisms of pattern control in the tetrapod limb (and vice versa).

The developmental perspective, and particularly structuralist approaches do, however, cause one to ask a whole new set of questions about the concept of adaptation as a process. In the traditional externalist view, an adaptation (as a functional structural feature) is thought to be shaped by selection and the cause of adaptation (both the process and the result) is largely contained within those external and independent factors. In an internalist mechanism the process of change is driven by internal factors and the process of selection is restricted to a 'purer' role in acceptance or rejection of the result. The difference is a subtle one, but it leads to complications. For example, is the demonstration of a developmental basis for a particular morphology (for example, in Gould's recent study of shell shape in *Cerion*, 1984) sufficient to represent a case of 'non-adaptation'?

This last point then leads to the important conclusion that a major gap in our knowledge of internal causes in evolutionary mechanisms is their environmental context. In addition to the genetic origins of particular changes, what are the ecological factors? How are the internal and external environments related? Again, this brings us back to Waddington and the whole question of genetic assimilation (Rachootin and Thomson 1981; Ho and Saunders 1979), and no doubt a range of far more complex phenomena. It also brings us to new questions about the inter-relationships of internal processes acting within development and externalist processes acting within populations. To what extent can selection reinforce the causation of change in developmental mechanisms?

Finally, the whole question of homology is opened up again, especially by the structuralist approach (see Goodwin 1984). Much of evolutionary biology (and indeed all comparative biology) depends upon systematic and phylogenetic analyses; and these in turn depend in great part upon knowing a given structure in one organisms is or is not the same at that in another. Traditionally, it has been sufficient to follow the 'history' of structures to make the case, the history within ontogeny being thought to mirror the history in phylogeny. However, now we can be less sure. In any scheme of pattern control, perhaps only the 'field' that is controlled is homologous across taxa. It is exciting that both in specialized questions such as homology and in the general question of cause in evolutionary change, we have finally and irrevocably moved from studying 'histories' to studying the causal mechanisms.

References

Beer, G. R., de (1940). *Embryos and Ancestors*. Clarendon Press, Oxford.
Bonner, J. T. (ed) (1982). *Evolution and Development*. Springer Verlag, Berlin, Heidelberg, New York.
Cooke, J. and Webber, J. A. (1983). Vertebrate embryos: diversity in developmental strategies. *Nature,* **306**, 423–4.
Galau, G. A., Klein, W. H., Davis, M. M., Wold, B. J., Britten, R. J., and Davidson, E. H. (1976). Structural genes sets active in embryos and adult tissues of the sea urchin. *Cell,* **7**, 487–505.
Goodwin, B. C. (1982). Development and Evolution. *J. theoret. Biol.* **85**, 757–70.
—— (1984). Changing from an evolutionary to a generative paradigm in biology. In *Evolutionary Theory* (ed. J. W. Pollard), pp. 99–119. John Wiley, New York.
—— and Trainor, J. E. H. (1980). A field description of the cleavage process in

embryogenesis. In *Development and Evolution* (eds B. C. Goodwin, N. Holder, and C. C. Wylie), pp. 75–98. Cambridge University Press.

——, Holder, N., and Wylie, C. C. (eds) (1983). *Development and Evolution*. Cambridge University Press, Cambridge.

Gould, S. J. (1977). *Ontogeny and Phylogeny*. Harvard University Press, Cambridge, Mass.

—— (1984). Morphological channeling by structural constraint. *Paleobiol.* **10**, 172–94.

Graham, C. F. and Wareing, P. F., (eds) (1984). *Developmental Control in Animals and Plants* (2nd edn). Blackwell Scientific Publications, Oxford.

Hanken, J. (1983). High incidence of limb skeletal variants in a peripheral population of the red-backed salamander *Plethodon cinerus* (Amphibia, Plethodontidae), from Nova Scotia. *Can. J. Zool.* **61**, 1925–31.

Herschel, J. F. W. (1830). *A Preliminary Discourse on the Study of Natural Philosophy*. Longmans, London.

Hinchliffe, J. R. and Johnson, D. R. (1982). *The Development of the Vertebrate Limb*. Clarendon Press, Oxford.

Ho, M.-W. and Saunders, P. T. (1979). Beyond neo-Darwinism: an epigenetic approach to evolution. *J. theoret. Biol.* **78**, 573–91.

Hodge, M. J. S. (1977). The structure and strategy of Darwin's 'Long Argument'. *Br. J. Hist. Sci.* **10**, 237–46.

Holder, N. (1983). Developmental constraints and the evolution of vertebrate digit patterns. *J. theoret. Biol.* **104**, 451–71.

Hull, D. L. (1980). Individuality and selection. *Ann. Rev. Ecol. Syst.* **11**, 311–32.

Lewontin, R. C. (1970). The units of selection. *Ann. Rev. Ecol. Syst.* **1**, 1–18.

Murray, J. D. (1981). A pre-pattern formation mechanism for animal coat markings. *J. theoret. Biol.* **88**, 161–99.

Odell, G., Oster, G. F. and Alberch, P. (1980). Mechanisms, morphogenesis and evolution. In *Lectures on Mathematics in the Life Sciences* (ed. G. Oster) pp. 165–255. American Mathematics Society, Providence, R.I.

Rachootin, S. P. and Thomson, K. S. (1981). Epigenetics, palaeontology and evolution. In *Evolution Today* (eds G. G. E. Scudder and C. L. Reveal), pp. 181–93. Hunt Institute, Pittsburg.

Raff, R. A. and Kaufman, T. C. (1983). *Embryos, Genes and Evolution*. Macmillan, New York.

Riedl, R. (1978). *Order in Living Organisms*. Wiley, New York.

Salthe, S. (1985). *Evolving Hierarchical Systems*. Columbia University Press, New York. (in press).

Seilacher, A. (1970). Arbeitskonzept zur Konstruktionsmorphologie. *Lethaia*, **3**, 393–6.

Slack, J. M. W. (1983). *From Egg to Embryo*. Cambridge University Press, Cambridge.

Stanley, S. M. (1975). A theory of evolution above the species level. *Proc. Nat. Acad. Sci. USA,* **72**, 646–70.

Thomson, K. S. (1982). The meanings of evolution. *Am. Scient.* **70**, 529–31.

—— (1984). Reductionism and other -isms in biology. *Am. Scient.,* **72**, 388–90.

Vrba, E. S. and Eldredge, N. (1984). Individuals, hierarchies and processes: towards a more complete evolutionary theory. *Paleobiol.* **10**, 146–72.

Waddington, C. H. (1975). *The Evolution of an Evolutionist*. Cornell University Press, Ithaca, New York.

Webster, G. C. and Goodwin, B. C. (1982). The origin of species: a structuralist approach. *J. soc. Biol. Struct.* **5**, 15–47.

Williams, G. C. (1985). A defence of reductionism in evolutionary biology. *Oxford Surv. Evolut. Biol.* **2**, 1–27.

Index

OXFORD SURVEYS IN EVOLUTIONARY BIOLOGY

Volume 1: 1984

Contents

Oxford Surveys in Evolutionary Biology

Oxford Surveys in Evolutionary Biology may be obtained from your local bookseller or on direct subscription through the Journals Subscription Department at Oxford University Press.

Once recorded as a direct subscriber, you will receive a renewal notice some six to eight weeks prior to the publication of each volume, with details of its content. This will allow plenty of time to renew your subscription and have **Oxford Surveys in Evolutionary Biology** delivered directly to you immediately on publication.

Simply fill in the reservation form overleaf and return it to the Journals Subscription Department, Oxford University Press, Walton Street, Oxford OX2 6DP, UK. Back volumes may also be ordered on this form.

We regret, it is not possible to dispatch any volumes until payment has been received.

RESERVATION FORM *Please print*

I have purchased Volume 2 of **Oxford Surveys in Evolutionary Biology** and I wish to be recorded as a direct subscriber and to receive regular renewal notices.

Name ....

Address ....

....

Postcode .... Country ....

Please supply the following back volumes of **Oxford Surveys in Evolutionary Biology**

Volume 1, 1984 £25.00 UK, $50.00 N. America, £32.00 Elsewhere

I enclose my remittance of ....
Please debit my Access/American Express/Diners/Visa Account*

Card Number ☐☐☐☐☐☐☐☐☐☐☐☐☐☐☐☐☐☐

Expiry Date .... Signature ....
If address registered with card company differs from above, please give details
Delete as applicable

It is also possible to subscribe to the following Oxford Annuals:
Oxford Reviews of Reproductive Biology
Oxford Surveys on Eukaryotic Genes
Oxford Surveys of Plant Molecular & Cell Biology